Politics in Mexico

Politics in Mexico

RODERIC AI CAMP

New York   Oxford
OXFORD UNIVERSITY PRESS
1993

To My Teachers

Oxford University Press

Oxford   New York   Toronto
Delhi   Bombay   Calcutta   Madras   Karachi
Kuala Lumpur   Singapore   Hong Kong   Tokyo
Nairobi   Dar es Salaam   Cape Town
Melbourne   Auckland

and associated companies in
Berlin   Ibadan

Copyright © 1993 by Oxford University Press, Inc.

Published by Oxford University Press, Inc.,
200 Madison Avenue, New York, New York 10016

Oxford is a registered trademark of Oxford University Press

Library of Congress Cataloging-in-Publication Data
Camp, Roderic Ai. Politics in Mexico / Roderic Ai Camp.
p. cm.   Includes bibliographical references and index.
ISBN 0-19-508074-2
ISBN 0-19-507612-5 (pbk)
1. Mexico—Politics and government.   I. Title.
JL1281.C35   1993   320.972—dc20   92-27455

1 3 5 7 9 8 6 4 2

Printed in the United States of America
on acid-free paper

Acknowledgments

Anyone who has been in the business of teaching eventually writes a mental textbook, constantly revised and presented orally in a series of lectures. As teachers, however, we often dream of writing just the right book for our special interest or course. Such a book naturally incorporates our own biases and objectives. It also builds on the knowledge and experiences of dozens of other teachers. While still a teenager, I thought of being a teacher, and perhaps unusually, a college professor. Teachers throughout my life, at all levels of my education, influenced this choice. They have also affected the way in which I have taught, my interrelationship with students, and my philosophy of learning and life. To these varied influences, I offer heartfelt thanks, and hope that this work, in some small way, repays their contributions to me personally and professionally, and to generations of other students.

Among those special teachers, I want to mention Thelma Roberts and Helen Weishaupt, who devoted their lives to the betterment of young children, instilling worthy values and beliefs and setting admirable personal examples, and to Mrs. Lloyd, for numerous afternoon conversations at Cambridge School. I wish to thank Ralph Corder and Don Fallis, who encouraged my natural interest in history toward a more specific interest in social studies. Sharon Williams and Richard W. Gully, my toughest high school teachers, introduced me to serious research, and the joys of investigating intellectual issues; and Inez Fallis, through four years of Spanish, prompted my continued interest in Mexico. Robert V. Edwards and Katharine Blair stressed the importance of communication, orally and in writing, helping me understand essential ingredients in the process of instruction. My most challenging professor, Dr. Bergel, during a high school program at Chapman College, opened my eyes to Western civilization, and to the intellectual feast that broad interdisciplinary teaching could offer.

For his humanity, advice, and skill with the English language, I remain indebted to George Landon. As a mentor in the classroom and a model researcher, Mario Rodríguez led me to the Library of Congress, and to the joys of archival research. On my arrival in Arizona, Paul Kelso took me under his wing, contributing vastly to my knowledge of Mexico and the out-of-doors, sharing a rewarding social life with his wife, Ruth. I learned more about Latin America and teaching in the demanding classroom setting of George A. Brubaker and Edward J. Williams. Both instilled the importance of clarity, teaching writing as well as substance. Finally, Charles O. Jones and Clifton Wilson set examples in their seminars of what I hoped to achieve as an instructor.

Indirectly, I owe thanks to hundreds of students who have graced my classrooms and responded enthusiastically, sometimes less so, to my interpretations of Mexican politics. I am equally indebted to Bill Beezely, David Dent, Oscar Martínez, and Edward J. Williams, devoted teachers and scholars, who offered many helpful suggestions for this book.

Contents

Politics in Mexico

1

Mexico in Comparative Context

> Neither this authoritarian interpretation nor the democratic interpretation which preceded it sufficiently explain Mexican politics. Although the authoritarian model seems more defensible, a more thorough reexamination of the political system is needed. Such a reexamination must be grounded in the roots of contemporary Mexico. More scholars need to follow Arnaldo Córdoba's lead and look to the Revolution for answers. . . . What emerged was a leviathan state capable of controlling society.
>
> DONALD J. MABRY, "Changing Models of Mexican Politics"

An exploration of a society's politics is by nature all-encompassing. Political behavior and political processes are a reflection of a culture's evolution, involving history, geography, values, ethnicity, religion, internal and external relationships, and much more. As social scientists, we often pursue the strategies of the modern journalist in our attempt to understand the political news of the moment, ignoring the medley of influences from the past.

Naturally, each individual tends to examine another culture's characteristics, political or otherwise, from his or her own society's perspective. This is not only a product of one's ethnocentrism, thinking of your society as superior to the next person's, for which we Americans are often criticized, but also a question of familiarity. Although we often are woefully ignorant of our own society's political processes and institutions, being more familiar with the mythology than actual practice, we become accustomed to our way of doing things in our own country.[1]

I will attempt to explain Mexican politics, building on this natural proclivity to relate most comfortably to our own political customs, by drawing on implied as well as explicit comparisons to the United States. We are also products of a more comprehensive Western European civilization, into which other traditions are gradually making significant inroads. Although some critics suggest that we have relied too exclusively on Western traditions in our education, they are unquestionably the primary source of our political values. Thus, our familiarity with political processes, if it extends at all beyond United States boundaries, is typically that of the western European nations and England.[2] For recent immigrants, of course, that heritage is different. Again, where possible, comparisons will be drawn to some of these political systems in order to place the Mexican experience in a larger context. Finally, Mexico is a Third World country, a category into which most countries fall, and hence its characteristics deserve to be compared to characteristics we might encounter elsewhere in the Third World.

WHY COMPARE POLITICAL CULTURES?

The comparison of political systems is an exciting enterprise. One reason that the study of politics in different societies and time periods has intrigued inquiring minds for generations is the central question: Which political system is best? Identifying the "best" political system, other than its merely being the one with which you are most familiar and consequently comfortable, is, of course, a subjective task. It depends largely on what you want out of your political system. The demands made on a political system and its ability to respond efficiently and appropriately to them are one way of measuring its effectiveness.

Throughout the twentieth century perhaps the major issue attracting the social scientist, the statesperson, and the average, educated citizen is which political system contributes most positively to economic growth and societal development. From an ideological perspective, much of international politics since World War II has focused primarily on that issue. As Peter Klarén concluded, "U.S. policymakers searched for arguments to counter Soviet claims that Marxism represented a better alternative for development in the Third World than did Western capitalism. At the same time U.S. scholars began to study in earnest the causes of underdevelopment. In particular scholars asked why the West had developed and why most of the rest of the world had not."[3]

The two political systems most heavily analyzed since 1945 have been democratic capitalism and Soviet-style socialism. Each has its pluses and minuses, depending on individual values and perspectives. Given recent events in eastern Europe and the breakup of the Soviet state, socialism is in decline. Nevertheless, socialism as a model is not yet dead, nor is it likely to be in the future. Administrators of the socialist model rather than the weaknesses inherent in the ideology can always be blamed for its failures. Furthermore, it is human nature to want alternative choices in every facet of life. Politics is just one facet, even if somewhat all-encompassing. The history of humankind reveals a continual competition between alternative political models.

In short, whether one chooses democratic capitalism, a fresh version of socialism, or some other hybrid ideological alternative, societies and citizens will continue to search for the most viable political processes to bring about economic and social benefits. Because most of the earth's peoples are economically underprivileged, they want immediate results. Often, politicians from less fortunate nations seek a solution through emulating wealthier (First World) nations. Mexico's leaders and its populace are no exception to this general pattern.

One of the major issues facing Mexico's leaders is the nature of its capitalist model, and the degree to which Mexico should pursue a strategy of economic development patterned after that of the United States. Since 1988 they have sought to alter many traditional relationships between government and the private sector, increasing the influence of the private sector in an attempt to reverse Mexico's economic crisis and stimulate economic growth. In fact, Mexico received international notice in the 1990s for the level and pace of change under President Carlos Salinas de Gortari.[4]

Salinas in public statements and political rhetoric has called for economic and political modernization. He has explicitly incorporated political with economic change, even implying a linkage.[5] Thus, he advocates economic liberalization, which he defines as increased control of the economy by the private sector; more extensive foreign investment; and internationalization of the Mexican economy through expanded trade and formal commercial relationships with the United States and Canada. Simultaneously, Salinas advocates political liberalization, which he defines as including more citizen participation in elections, greater electoral competition, and integrity in the voting process—all features associated with the United States and European liberal political traditions.

It is hotly debated among social scientists whether a society's political model determines its economic success or whether its economic model produces its political characteristics. Whether capitalism affects the behavior of a political model or a political model is essential to successful capitalism leads to the classic chicken-and-egg argument. It may well be a moot point because the processes are interrelated in terms of not only institutional patterns but cultural patterns as well.[6]

The comparative study of politics reveals, to some extent, a more

important consideration. If the average Mexican is asked to choose between more political freedom or greater economic growth, as it affects him
or her personally, the typical choice is the latter.[7] This is true in other Third
World countries too. People with inadequate incomes are much more likely
to worry about bread-and-butter issues than about more political freedom.
A country's political model becomes paramount, however, when its citizens draw a connection between economic growth (as related to improving
their own standard of living), and the political system. If they believe the
political system, and not just the leadership itself, is largely responsible for
economic development, it will have important repercussions on their political values and their political behavior. If Mexicans draw such a connection,
it will change the nature of their demands on the political leadership and
system, and the level and intensity of their participation.[8]

The comparative study of societies provides a framework by which
we can measure the advantages and disadvantages of political models as
they impact on economic growth. Of course, economic growth itself is not
the only differentiating consequence. Some political leaders are equally
concerned, in some cases more concerned, with social justice. Social
justice may be interpreted in numerous ways. One way is to think of it as a
means of redistributing wealth. For example, we often assume that economic growth, the percentage by which a society's economic productivity
expands in a given time period, automatically conveys equal benefits to
each member of the society. More attention is paid to the level of growth,
than to its beneficiaries. It is frequently the case that the lowest-income

---

Social justice: a concept focusing on each citizen's quality of life and
the equal treatment of all citizens.

---

groups benefit least from economic growth. This has been true in the
United States, but is even more noticeable in Third World and Latin American countries. There are periods, of course, when economic growth produces greater equality in income distribution.[9] Per capita income figures
(national income divided by total population) can be deceiving because
they are averages. In Mexico, for example, even during the sustained
growth of the 1950s and 1960s, the real purchasing power (ability to buy
goods and services) of the working classes actually declined.[10] Higher-
income groups increased their proportion of national income from the
1970s to the 1990s, decades of economic crisis, and that of the lower-
income groups declined.[11]

Another way of interpreting social justice is on the basis of social
equality. This does not mean that all people are equal in ability but that

each individual should be treated equally under the law. Social justice also implies a leveling of differences in opportunities to succeed, giving each person equal access to society's resources. Accordingly, its allocation of resources can be a measure of a political system.

The degree to which a political system protects the rights of all citizens is another criterion by which political models can be compared. In Mexico, where human rights abuses are a serious problem, the evidence is unequivocal that the poor are much more likely to be the victims than are members of the middle and upper classes. The same can be said about many societies, but sharp differences in degree exist between highly industrialized nations and Third World nations.[12]

From a comparative perspective, then, we may want to test the abilities of political systems to eliminate both economic and social inequalities. It is logical to believe that among the political models where the population has a significant voice in making decisions, the people across the board obtain a larger share of the societal resources. On the other hand, it is possible to argue, as in the case of Cuba, that an authoritarian model can impose more widespread, immediate equality in the distribution of resources, even in the absence of economic growth, while reducing the standard of living for formerly favored groups.

Regarding social justice and its relationship to various political models, leaders are also concerned with the distribution of wealth and resources *among* nations, not just within an individual nation. The choice of a political model, therefore, often involves international considerations. Such considerations are particularly important to countries that have achieved independence in the twentieth century, especially after 1945. These countries want to achieve not only economic but also political and cultural independence. Although Mexico, like most of Latin America, achieved political independence in the early nineteenth century, it found itself in the shadow of an extremely powerful neighbor. Its proximity to the United States eventually led to its losing half of its territory and many natural resources.

A third means to compare political models is ability to remake a citizenry. A problem faced by most nations, especially in their infancy, is to build a sense of nationalism. A sense of nationalism is difficult to erase, even after years of domination by another power, as in the case of the Soviet Union and the Baltic republics, but is equally difficult to establish, especially in societies incorporating diverse cultural, ethnic, and religious heritages.[13] The political process can be used to mold citizens, to bring about a strong sense of national unity, while lessening or dampening local and regional loyalties. The acceptability of a political model, its very legitimacy among the citizenry, is a measure of its effectiveness in developing national sensibilities. Mexico, which had an abiding sense of regionalism, struggled for many decades to achieve a strong sense of national unity and pride.[14] On the other hand, Mexico did not have sharp religious and ethnic differences, characteristic of other cultures, such as India, to overcome.

Many scholars have suggested that the single most important issue governing relationships among nations in the twenty-first century will be that of the haves versus the have nots.[15] In fact, the probability that Mexico might be linked to the United States and Canada in a free trade agreement highlights the point. One of the arguments against such an agreement is the impossibility of eliminating trade barriers between a nation whose per capita wage is one-seventh of the per capita wage of the other nation.[16] One of the arguments for such an agreement is that it could temper the disparity.

The dichotomy between rich and poor nations is likely to produce immense tensions in the future, yet the problems both sets of nations face are remarkably similar. As the 1990 World Values Survey illustrates, an extraordinary movement in the coincidence of some national values is afoot, for example, in the realm of ecology. This survey, which covers forty countries, discovered that from 1981 to 1990 an enormous change in concern about environmental issues occurred in poor as well as rich nations. Other problems that most countries, regardless of their standard of living or political system share include availability of natural resources, notably energy; production of foodstuffs, especially grains; level of inflation; size of national debt; access to social services, including health care; inadequate housing; and maldistribution of wealth.

Another reason that examining political systems from a comparative perspective is useful is personal. As a student of other cultures you can learn more about your own political system by reexamining attitudes and practices long taken for granted. In the same way a student of foreign languages comes to appreciate more clearly the syntax and structure of his or her native tongue, and the incursions of other languages in its constructions and meanings, so too does the student of political systems gain. Comparisons not only enhance your knowledge of the political system within which you live but are likely to increase your appreciation of selected features.

Examining a culture's politics implicitly delves into its values and attitudes. As we move quickly into an increasingly interdependent world, knowledge of other cultures is essential to being well educated. Comparative knowledge, however, allows you to test your values against those of other cultures. How do yours measure up? Do other sets of values have applicability in your society? Are they more or less appropriate to your society? Why? For example, one of the reasons that considerable misunderstanding exists between the United States and Mexico is a differing view of the meaning of political democracy. Many Mexicans attach features to the word *democracy* that are not attached to its definition in the United States.[17] For example, as will be discussed later, for many Mexicans, democracy does not incorporate tolerance for opposing viewpoints. Problems arise when people do not realize they are using a different vocabulary when discussing the same issue.

Another reason for striking out on the path of comparing political cultures is to dispel the notion that Western, industrialized nations have all of the solutions. It is natural to think of the exchange of ideas favoring the most technologically developed nations, including Japan, Germany, and the United States. Solutions do not rely on technologies alone; in fact, most rely on human skills. In other words, how do people do things? This is true whether we are analyzing politics or increasing sales in the marketplace. Technologies can improve efficiency, quality, and output of goods and services, yet their application raises critical human issues revolving around values, attitudes, and interpersonal relationships. For example, the Japanese have a management philosophy governing employee and employer relations. It has nothing to do with technology. Many observers believe, however, that the philosophy in operation produces better human relationships and higher economic productivity. Accordingly, it is touted as an alternative model in the workplace. The broader the scope of human understanding, the greater our potential for identifying and solving human-made problems.

Finally, as a student new to the study of other cultures, you may be least interested in the long-term contributions such knowledge can make for its own sake. Yet, our ability to explain differences and similarities between and among political systems and, more important, their consequences, is essential to the growth of political knowledge. Although not always the case, it is generally true that the more you know about something and the more you understand its behavior, the more you can explain its behavior. This type of knowledge allows social scientists to create new theories of politics and political behavior, some of which can be applied to your own political system as well as to other cultures. It also allows, keeping in mind the limitations of human behavior, some level of prediction. In other words, given certain types of institutions and specific political conditions, social scientists can predict that political behavior is likely to follow certain patterns.

SOME INTERPRETATIONS OF THE MEXICAN SYSTEM

We suggested above that social scientists set for themselves the task of formulating some broad questions about the nature of a political system and its political processes. A variety of acceptable approaches exist from which to examine political systems individually or comparatively. Some approaches tend to focus on relationships among political institutions and the functions each institution performs. Other approaches give greater weight to societal values and attitudes, and the consequences these have for political behavior and the institutional features characterizing a political system. Still other approaches, especially in the last third of the twentieth century, place greater emphasis on economic relationships and the influence of social or income groups on political decisions. Taking this last approach a step further, many analysts of Third World countries, including Mexico, focus on international economic influences and their effect on domestic political structures.

Choosing any one approach to explain the nature of political behavior has advantages once you undertake to describe a political system. In my own experience, however, I have never become convinced that one approach offers an adequate explanation. I believe, however, that an examination of political processes or functions entails the fewest prejudices, and that by pursuing how and where these functions occur, one uncovers the contributions of other approaches.[18] An eclectic approach to politics, incorporating culture, history, geography, and external relations, provides the most adequate and accurate vision of contemporary political behavior. Such an eclectic approach, combining the advantages of each, will be used in this book.

In general, the study of Mexican politics has provoked continued debate about which features impact most on political behavior, and more commonly, to what degree is Mexico an authoritarian model.[19] It is important, as we begin this exploratory task, to offer some explanations about the nature of Mexican politics.

Does Mexico have an authoritarian political system? The answer simply is yes. Is Mexican politics in the same authoritarian category as Cuba under Fidel Castro, China since the Communist Revolution, or the Soviet Union before 1991? The answer definitely is no. Mexico can best be described as a semiauthoritarian political system—a hybrid of political liberalism and authoritarianism that gives it a special quality or flavor—

that is well documented institutionally in the 1917 Constitution, currently in effect.

Mexico's unique authoritarianism sets it apart from many other societies, including Latin American countries that have passed through long periods of authoritarian control, especially in the 1970s and 1980s.[20] Normally, *authoritarian* refers to a political system in which fewer individuals have access to the decision-making process, and fewer still are in a position to exercise important political choices and policies.

What sets Mexico's authoritarian system apart from many others is that it allows much greater access to the decision-making process and, more important, its decision makers change frequently.[21] Usually, the advantage of a well-established authoritarian regime is continuity. Whereas it is fair to say that successive generations of Mexican leaders, with ties to

---

Authoritarian: in political terms, a system in which only a small number of individuals exercise and have access to political power.

---

their predecessors, have controlled the decision-making process, this has not led necessarily to continuity in policy. Furthermore, its leadership, in the hands of the executive branch, especially the president, is limited to a six-year term.

A second feature of the Mexican political model, integral to its hybrid authoritarianism, is a special feature found in many Latin American cultures: *corporatism.* Corporatism in this political context refers to how groups in society relate to the government, or more broadly the state; the process through which they channel their demands to the government; and how the government responds to their demands.[22] In the United States, any introductory course in U.S. politics devotes some time to interest groups, and how they present their demands to the political system. Mexico, which inherited the concept of corporatism from Spain, instituted a corporate relationship between the state and various important interest or social groups in the 1930s, primarily under the presidency of General Lázaro

---

Corporatism: the more formal relationship between selected groups or institutions, and the government or state.

Cárdenas (1934–1940). This means that the government took the initiative to strengthen various groups, creating umbrella organizations to house them and through which their demands could be presented. The government placed itself in an advantageous position by representing various interest groups, especially those most likely to support opposing points of view. The state attempted, and succeeded over a period of years, in acting as the official arbiter of these interests. It generally managed to make various groups loyal to it in return for representing their interests. For example, it absorbed the largest groups in a government-sponsored political party, now called the Institutional Revolutionary Party (Partido Revoluciónario Institucional—the PRI), giving them legitimacy and a role in party affairs. These included peasants, labor, and middle-class professional groups.

The essence of the corporatist relationship is political reciprocity. In return for official recognition and official association with the government or government-controlled organizations, these groups could expect some consideration of their interests on the part of the state. They could also expect the state to protect them from their natural political enemies. For example, labor unions hoped the state would favor their interests over the interests of powerful businesses.

The corporatist structure has led to a situation in which the state is the all-powerful force in the society, and it is often patronizing in its relations with various groups. Corporatism facilitates the state's ability to manipulate various groups in the state's own interest. In other words, Mexican political leadership itself might be thought of as a separate interest group, but unlike all other interest groups, it is in control of the decision-making process.[23] The state's uncontested dominance has led to its leaders making some choices that benefit themselves rather than various social-class or group interests.

Mexico's political system is not only semiauthoritarian and corporatist but allows the government and/or state to play a paramount role, a third distinguishing feature. State institutions have generally had far more prestige, resources, and influence than private, independent, or nonprofit organizations have had. The state's prestige has contributed to the perpetuation of its influence. Many of the best minds, regardless of profession or educational background, are attracted to lifetime careers with the state.[24] (This is also true elsewhere in Latin America, and throughout Asia and Africa.) State dominance has contributed enormously to the growth and centralization of resources in the capital city and the Federal District, somewhat analogous to the District of Columbia in the United States. The power of the Mexican national state and the comparative weakness of local and provincial authorities contribute to a mentality of dependence on the state. The excessive dependence has engendered resentment as well, as various groups and geographic regions have sought to establish their autonomy from centralized state control.[25]

The dominance of the state within a semiauthoritarian, corporatist political structure, contributes to a fourth political feature of the Mexican model: the centralization of authority in the executive branch. The Mexican model is unquestionably presidentially dominant, a phenomenon that Mexicans refer to as *presidencialismo* (presidentialism).[26] Americans think of their president as being tremendously influential. Of course, no other individual citizen or official can exercise the level of political influence that the U.S. president can. His influence is further exaggerated because he is seen as the, or a, major world leader. The Mexican president has no comparable international credentials—even though President Salinas (1988–1994) enhanced his domestic status considerably by virtue of

---

Presidencialismo: the concept that most political power lies in the hands of the president, and all that is good or bad in government policy stems personally from the president.

---

his international reputation—but the president nevertheless exercises far more control over the Mexican political scene than does his American counterpart in the United States. The strength of the presidency specifically, and the executive branch generally, is at the cost of an ineffectual legislative and judicial branch, or any other autonomous authority.

The importance of Mexican executive leadership and the dominance of the state has led to the development of a dynamic political elite whose careers are formed within the national governmental bureaucracy. The elite, which has never been characterized by ideological homogeneity, is relatively open, but recently has taken on fairly homogeneous social, career, and educational characteristics.[27] Although entry into political leadership ranks is available to well-educated Mexicans, it is a self-designated group. That is to say, most important decision-making posts are appointive in nature, and those selecting the officeholders are themselves incumbents. These individuals change over time, thus facilitating access to leadership positions and the alteration of policy goals, but they are not as responsive to constituencies as are U.S. officials.[28] Moreover, because a primary goal of politicians everywhere is to stay in power, Mexican politicians, lacking constituent responsibilities, have generally been pragmatic, doing whatever is necessary to remain in office rather than pursuing a committed, ideological platform.

The structural features of Mexico's political model—semiauthoritarianism, corporatism, state dominance, centralization of authority, and a self-selecting elite—are complemented by a dual political heritage incorporated into the political culture. The political culture is neither democratic nor authoritarian. It is contradictory: modern and traditional. Mexico, as Noble Prize–winner Octavio Paz argues, is built from two different populations, rural versus urban, traditional versus modern.[28] It labors under many historical experiences, precolonial, colonial, independence, and revolutionary. The experiences have led to a political culture that admires essential democratic values, such as citizen participation, yet strongly favors intolerance of opposing points of view.[30] It is the cultural blend of contradictory values that explains Mexico's special authoritarian system. The contradictions in its political culture and historical experiences have also produced a set of policy goals, many incorporated in the Constitution, that too are contradictory. On one hand, a strong state is favored; on the other, capitalism is the preferred tool for economic growth.

Place and historical experience have also contributed to another feature of mass political culture: a dependent psychology.[31] The proximity of the United States, which shares a border with Mexico nearly two thousand miles long, and the extreme disparities between the two in economic wealth and size tend to foster an inferiority complex in many Mexicans, whether they operate in the worlds of business, academia, technology, or politics. The economic, cultural, and artistic penetration of the United States into Mexico carries with it other values foreign to its domestic political heritage. Psychologically and culturally, Mexicans must cope with these influences, most of which are indirect, often invisible. A strong sense of Mexican nationalism, especially in relation to its political model, is expressed in part as a defensive mechanism against United States influences. This level of nationalism has produced and sustained unique characteristics of the Mexican political model.

MEXICO'S SIGNIFICANCE IN A COMPARATIVE CONTEXT

From a comparative perspective, Mexico provides many valuable insights into politics and political behavior. The feature of Mexico that has most intrigued students of comparative politics is the stability of its political system.[32] Although challenged seriously by military and civilian factions in 1923, 1927, and 1929, its political structure and leadership have obtained for most of this century, at least since 1930—an accomplishment unmatched by any other Third World country. Even among industrialized nations like Italy, Germany, and Japan, such longevity is remarkable. The phenomenon evokes questions. What enables the stability? What makes the Mexican model unique? Is it the structure of the model? Is it the political culture? Does it have something to do with the country's proximity to another exemplar of political continuity? Or with the values and behavior of the people?

We know from other studies of political stability that a degree of political legitimacy accompanies even a modicum of support for a political model. Although social scientists are interested in political legitimacy and political stability each for its own sake, they assume, with considerable evidence, that some relationship exists between economic development and political stability. Although it is misleading to think that the characteristics of one system can be successfully transferred to another, it is useful to ascertain which may be more or less relevant to accomplishing specific, political goals.

Mexico has also attracted considerable international interest because it is a one-party-dominant system encountering only limited opposition from 1929 through 1988, the year in which a splinter group from the official party ran a highly successful campaign. Mexico's system is unusual in that the antecedent of the PRI, the National Revolutionary Party (Partido Nacional Revolucionario, the PNR), did not bring the political leadership to power. Rather, the leadership established the party as a vehicle to *remain* in power; the PRI was founded and controlled by the government bureaucracy. This had long-term effects on the nature of the party itself, and on its importance to policy-making.[33] In this sense, the PRI is unlike the Communist Party in the Soviet Union, whose death in 1991 spelled the end of Communist leadership in the successor states. The PRI, because it does not produce Mexico's leadership, as do the Democratic and Republican parties in the United States, is much more tangential to political power, and consequently much more expendable.

A third reason that Mexico's political system intrigues outside observers has been its ability to subordinate military authorities to civilian control. Mexico, like most other Latin American countries, passed through a century when violence became an accepted tool of the political game. Such acceptance makes it extremely difficult, if not impossible, to eliminate the military's large and often decisive political role. Witness many Latin American countries;[34] one has only to look at Argentina and Chile during the 1970s and 1980s. No country south of Mexico has achieved its extended *civilian* supremacy. In most Latin American countries where civilian leadership is once again in ascendancy, their holds are tenuous at best.

Mexico, therefore, is a unique case study in Third World civil-military relations. What produced civilian supremacy there? Is the condition found elsewhere? A confluence of circumstances and policies gradually succeeded in putting civilian control incrementally in place. Some involve the special characteristics of the system itself, including the creation of a national political party. Some are historical, most important of which is the Mexican Revolution of 1910, which led to the development of a popular army whose generals governed Mexico in the 1920s and 1930s, and who themselves initiated the concept of civilian control.[35]

A fourth reason for studying Mexico is the singular relationship it has developed with the dominant religious institution, the Catholic church. Throughout much of Latin America, the Catholic church has been one of the important corporate actors. For significant historical reasons in the nineteenth and twentieth centuries, Mexico's leadership suppressed and then isolated itself from the Catholic hierarchy, and even in some cases, the Catholic religion.[36] The Catholic Church has often played a political role in Latin American societies, and currently has the potential to exercise considerable political and social influence. A study of Church-state relations in Mexico offers a unique perspective on how the Church was removed from the corporatist structure, and the implications of this autonomy for a politically influential institution.

Finally, a fifth reason for examining Mexico in a comparative political context is the opportunity to view the impact of the United States, a First World country, on a Third World country. No comparable geographic relationship obtains anywhere else in the world: two countries that share a long border exhibit great disparities in wealth. Mexico provides not only a test case for those who view Latin America as dependent on external economic forces but an unparalleled opportunity to look at the possible *political* and *cultural* influences and consequences of a major power.[37] The relationship is not one-way but it is asymmetrical.[38] The United States exercises or potentially exercises more influence over Mexico than vice versa. This does not mean that Mexico is the passive partner. It, too, exercises influence. In many respects its influence is growing. Because of European civilization's influence on our culture, we have long studied the political models of England and the Continent. Our obsession with the Soviet Union exaggerated our focus on Europe. As Hispanic and other immigrant cohorts grow larger in the United States, our knowledge of the Mexican culture will become far more relevant for understanding *contemporary political behavior* in the United States than anything we might learn from contemporary Europe.

CONCLUSION

To summarize, then, approaching politics from a comparative perspective offers many rewards. It allows us to test political models against one another; it makes it possible for us to learn more about ourselves and our own political culture; it offers a means for examining the relationship between political and economic development and the distribution of wealth; and it identifies common interests of rich and poor nations, and what they do to solve their problems.

Mexico's political system has many features, and scholars have interpreted them in different ways. This book argues that the system is semiauthoritarian, corporatist, state dominated; is led by a bureaucratic elite and a centralized executive; is built upon a contradictory political culture that includes authoritarian and liberal qualities; and is affected psychologically and politically by its proximity to the United States. Mexico offers unique opportunities for comparative study because of its political continuity and stability, one-party-dominant system, civil-military relations, unique separation of Church and state, and its nearness to a powerful, wealthy neighbor.

In the next chapter, the importance for Mexico of time, place, and historical roots is examined in greater detail and contrasted with the experiences of other countries. Among these elements are its Spanish heritage, the role of the state, nineteenth-century liberalism and positivism, the Revolution, and U.S.-Mexican relations.

NOTES

1. Gabriel Almond and Sidney Verba, *The Civic Culture* (Boston: Little, Brown, 1965), 59.

2. Compare for example, the number of academic course offerings and textbooks available on Europe and European countries with those representing other, especially Third World, regions and societies.

3. Peter Klarén "Lost Promise: Explaining Latin American Underdevelopment," in *Promise of Development: Theories of Change in Latin America,* ed. Peter Klarén and Thomas J. Bossert (Boulder, Colo.: Westview Press, 1986), 8.

4. See for example, the glowing statement by the *Washington Post,* that

Salinas "has proved to be as radical in his own way as the revolutionaries who galloped over Mexico at the beginning of the century." May 17, 1991.

5. For Salinas's views in English, see the interview "A New Hope for the Hemisphere," *New Perspective Quarterly* 8 (Winter 1991): 8.

6. The clearest presentation of this argument, in brief form, can be found in Gabriel Almond, "Capitalism and Democracy," *PS*, 24 (September 1991): 467–73.

7. In the World Values Survey (a collaborative survey of forty countries in 1981 and again in 1990, available in data format from the University of Michigan, Ann Arbor. Ronald Inglehart, Institute for Social Research, directed the North American project), 1990, data from Mexico show approximately 60 percent of the population choosing economic growth as most important, compared to approximately 25 percent who selected increased political participation.

8. On a presidential level, most Mexicans have not yet made the connection, or if they have, it is not significant to their voting. See Jorge Domínguez and James McCann, "Whither the PRI? Explaining Voter Defection from Mexico's Ruling Party in the 1988 Presidential Elections" (Paper presented at the Western Political Science Association meeting, March 1991), 23–24.

9. It has been argued, as a general rule, that as countries achieve advanced industrial economies, greater economic equality will be achieved. See Samuel P. Huntington, *Political Order in Changing Societies* (New Haven: Yale University Press, 1968), 57. Also see Dan LaBotz's statement that real minimum wages for Mexicans declined 44 percent between 1977 and 1988. *Mask of Democracy: Labor Suppression in Mexico Today* (Boston: South End Press, 1992), 19.

10. Roger D. Hansen, *The Politics of Mexican Development* (Baltimore: Johns Hopkins University Press, 1971), especially section "Trends in Mexican Income Distribution," 72ff.

11. Sidney Weintraub, *A Marriage of Convenience: Relations between Mexico and the United States* (New York: Oxford University Press, 1990), 36.

12. Of course, this is true worldwide. Unfortunately, the problem *seems* less severe when these groups are the primary victims. Americas Watch, *Human Rights in Mexico: A Policy of Impunity* (New York: Human Rights Watch, 1990), 53.

13. Karl W. Deutsch, *Nationalism and Social Communication: An Inquiry into the Foundations of Nationality,* 2d ed. (Cambridge: MIT Press, 1966), 156ff.

14. Frederick Turner, *The Dynamics of Mexican Nationalism* (Chapel Hill: University of North Carolina Press, 1968).

15. The classic argument for this was presented by Barbara Ward, *The Rich Nations and the Poor Nations* (London: Hamilton, 1962).

16. Jeff Faux, "No: The Biggest Export Will Be U.S. Jobs," *Washington Post Weekly Edition,* May 13–19, 1991, 8.

17. See one commissioner's statement that this is a source of bilateral problems in the blue-ribbon Report of the Bilateral Commission on the Future of United States-Mexican Relations, *The Challenge of Interdependence* (Lanham, Md.: University Press of America, 1989), 237.

18. The most comprehensive explanation of various interpretations still is Carolyn Needleman and Martin Needleman, "Who Rules Mexico? A Critique of Some Current Views of the Mexican Political Process," *Journal of Politics* 31 (November 1969): 1011–34.

19. For example, Susan K. Purcell, "Decision-Making in an Authoritarian Regime: Theoretical Implications from a Mexican Case Study," *World Politics* 26 (October 1973), 28–54.

20. José Luis Reyna and Richard Weinert, eds., *Authoritarianism in Mexico* (New York: ISHI Press, 1977).

21. Lorenzo Meyer, one of Mexico's foremost independent commentators, expresses it this way: "Mexico's system does allow, however, for some limited political pluralism and a higher degree of institutionalization and accessibility than the authoritarian regimes that dominated Latin America's Southern Cone a decade ago." See his "Democratization of the PRI: Mission Impossible?" in *Mexico's Alternative Political Futures*, ed. Wayne A. Cornelius, Judith Gentleman, and Peter H. Smith (La Jolla: Center for U.S.-Mexican Studies, UCSD, 1989), 333.

22. Ruth Spalding, "The Mexican Variant of Corporatism," *Comparative Political Studies* 14 (July 1981): 139–61.

23. For evidence of this view, see John W. Sloan, "State Power and Its Limits: Corporatism in Mexico," *Inter-American Economic Affairs* 38, no. 4 (1984): 3–18.

24. For the state's relationship to professional development, see Peter Cleaves, *Professions and the State: The Mexican Case* (Tucson: University of Arizona Press, 1987).

25. Edward J. Williams, "The Resurgent North and Contemporary Mexican Regionalism," *Mexican Studies* 6 (Summer 1990): 299–323, makes a strong case for northern regionalism.

26. Edmundo González Llaca, "El presidencialismo o la personalización del poder," *Revista Mexicana de Ciencias Políticas* 21 (Abril–Junio 1975): 35–42, discusses this at some length.

27. For some of these, see my "The Political Technocrat in Mexico and the Survival of the Political System," *Latin American Research Review* 20, no. 1 (1985): 97–118.

28. For detailed, long-term patterns of continuity and turnover, see Peter H. Smith, *Labyrinths of Power: Political Recruitment in Twentieth Century Mexico* (Princeton: Princeton University Press, 1979), 159ff. For comparisons to other countries, see John D. Nagle, *System and Succession, the Social Bases of Political Elite Recruitment* (Austin: University of Texas Press, 1977), 23ff.

29. Octavio Paz, *The Other Mexico: A Critique of the Pyramid* (New York: Grove Press, 1972), 45. Paz noted the existence of "one fundamental characteristic of the contemporary situation: the existence of two Mexicos, one modern and the other underdeveloped. This duality is the result of the Revolution and the development that followed it: thus, it is the source of many hopes and, at the same time, of future threats."

30. Enrique Alduncín, *Los valores de los mexicanos* (Mexico: Fomento Cul-

tural Banamex, 1986); Enrique Alduncín, *Los valores de los mexicanos, México en tiempos de cambio*, Vol. 2 (Mexico: Fomento Cultural Banamex, 1991); Raúl Béjar Navarro, *El mexicano, aspectos culturales y psicosociales* (Mexico: UNAM, 1981). The most comprehensive work in English is Rogelio Díaz Guerrero, *Psychology of the Mexican* (Austin: University of Texas Press, 1975).

31. See Octavio Paz's classic, *The Labyrinth of Solitude: Life and Thought in Mexico* (New York: Grove Press, 1961).

32. For an overview of these issues, see Kevin Middlebrook's review essay "Dilemmas of Change in Mexican Politics," *World Politics* 41 (October 1988): 120–41; for predictions about the future, see *Mexico's Political Stability: The Next Five Years*, ed. Roderic Ai Camp (Boulder, Colo.: Westview Press, 1986).

33. See Dale Story, *The Mexican Ruling Party, Stability and Authority* (New York: Praeger, 1986), 9ff; John J. Bailey, *Governing Mexico: The Statecraft of Crisis Management* (New York: St. Martin's Press, 1988).

34. This is nicely explained in Gary Wynia, *The Politics of Latin American Development*, 3d ed. (Cambridge: Cambridge University Press, 1990), 28ff.

35. For greater detail about the causes, see my *Generals in the Palacio: The Military in Modern Mexico* (New York: Oxford University Press, 1992).

36. Karl Schmitt, "Church and State in Mexico: A Corporatist Relationship," *Americas* 40 (January 1984): 349–76.

37. For some examples of noneconomic variables, see Clark W. Reynolds and Carlos Tello, *U.S.-Mexico Relations: Economic and Social Aspects* (Stanford: Stanford University Press, 1983).

38. For various insights on this, from the points of view of an American and Mexican, see Robert A. Pastor and Jorge G. Castañeda, *Limits to Friendship: The United States and Mexico* (New York: Vintage Press, 1989).

2

Political–Historical Roots: The Impact of Time and Place

> The political life of all those states which during the early years
> of the last century arose upon the ruins of the Spanish Empire
> on the American mainland presents two common features. In
> all those states, constitutions of the most liberal and democratic
> character have been promulgated; in all, there have from time
> to time arisen dictators whose absolute power has been either
> frankly proclaimed or thinly veiled under constitutional forms.
> So frequently has such personal rule been established in many
> of the states that in them there has appeared to be an almost
> perpetual and complete contradiction between theory and prac-
> tice, between nominal and the actual systems of government.
>
> CECIL JANE, *Liberty and Despotism in Spanish America*

Understanding politics is not just knowing who gets what, where, when
and how, as Harold D. Lasswell declared in a classic statement years ago,
but also understanding the origins of why people behave the way they do.
Each culture is a product of its own heritage, traditions emerging from
historical experiences. Many aspects of the U.S. political system can be
traced to our English colonial experiences, our independence movement,
our westering frontier expansion and our immigrant origins. Mexico has
had a somewhat similar set of experiences, but the sources of the experiences and their specific characteristics were quite different.

THE SPANISH HERITAGE

Mexico's political heritage, unlike that of the United States, draws on two
important cultural foundations: European and indigenous. Although large

numbers of Indians remained unintegrated with the conquering culture in
New Spain, a vast integration process took place in most of central Mexico. Contrariwise, British settlers encountered numerous Native Americans
in their colonization of North America, but they rarely intermarried with
these native peoples and thus the two cultures never blended. Racially,
African blacks played an important role in some regions; politically, a
limited role because of the small numbers brought to New Spain, the
colonial Spanish viceroyalty that extended from Central America to what is
now the United States Midwest and Pacific Northwest.

Mexico's racial heritage, unlike that of the United States, took on a
mixed or *mestizo* quality. The original Spanish conquerors, in the initial
absence of Spanish women, sought native mistresses or wives. In fact,
cohabiting with female royalty from the various indigenous cultures was
seen as an effective means of joining the two sets of leaders, firmly establishing Spanish ascendancy throughout the colony. The Indian-Spanish
offspring of these unions at first were considered socially inferior to Spaniards fresh from Spain and the Spanish born in the New World. Frank
Tannenbaum describes the complex social ladder:

> With the mixture of races in Mexico added to by the bringing in of Negroes
> in sufficient numbers to leave their mark upon the population in certain parts
> of the country, we have the basis of the social structure that characterized
> Mexico throughout the colonial period and in some degree continues to this
> day. The Spaniard—that is, the born European—was at the top in politics, in
> the Church, and in prestige. The *criollo,* his American-born child, stood at a
> lower level. He inherited most of the wealth, but was denied any important
> role in political administration. The *mestizo* and the dozen different *castas*
> that resulted from the mixtures of European, Indian, and Negro in their
> various degrees and kinds were still lower.[1]

In the late nineteenth century, mestizos reached a new level of social
ascendancy through their numbers and control over the political system.

Early Mexican political history involved social conflicts based on
racial heritage. Moreover, large indigenous groups were suppressed, ex-
ploited, and politically ignored. The prejudice with which Indians were
treated by the Spanish and mestizo populations, and the mistreatment of
the mestizo by the Spanish contributed further to the sharp class distinctions that have plagued Mexico.[2] Social prejudice was transferred to economic status as well: those lowest on the racial scale ended up at the
bottom of the economic scale. The degree of social inequality ultimately
contributed to the independence movement, as the New World–born Spanish (*criollos*) came to resent their second-class status relative to the Old
World–born Spanish (*peninsulares*). It contributed even more significantly

to the Mexican Revolution of 1910, in which thousands of downtrodden mestizo peasants and workers, and some Indians, joined a broad social movement for greater social justice.

All societies have some type of social structure. Most large societies develop hierarchical social groups, but from one society to another the level of deference exacted or given varies. In the United States, where political rhetoric, beginning with independence, focused on greater social equality, class distinctions were fewer and less distinct.[3] In Mexico, in spite of its Revolution, the distinctions remain much sharper, affecting various aspects of cultural and political behavior. For example, a major study of U.S. intellectuals found that 40 percent of the younger generation were from working-class backgrounds. By contrast, in Mexico, fewer than 5 percent fell into this social category.[4] In the political realm, lower-income groups, who are formally well-represented in the official government party, the PRI, are little represented in policy or leadership roles.[5] Lower-income groups have limited protection from abuses by governmental authorities and do not always receive equal treatment under the law. In the United States some differences exist in the legal treatment of rich and poor, but they are fewer, and the gap between them is much smaller than in Mexico.

The Spanish also left Mexico with a significant religious heritage: Catholicism. Religion played a critical role in the pre-Conquest Mexican indigenous culture, and was very much integrated into the native political processes. In both the Aztec and Maya empires, for example, religion was integral to political leadership. The Spanish were no less religious. Beginning with the Conquest itself, the pope reached some agreements with the Spanish crown. In these agreements, known collectively as the *patronato real* (royal patronate), the Catholic Church gave up certain rights it exercised in Europe for a privileged role generally in the Conquest and in New Spain specifically. In return for being allowed to send two priests or friars with every land or sea expedition, and being given the *sole* opportunity to proselytize millions of Indians, the Church gave up its control over the building of facilities in the New World, the appointing of higher clergy, the collecting of tithes, and other activities. In other words, Catholicism obtained a monopoly in the Spanish New World.[6]

The contractual relationship between the Catholic Church and the Spanish authorities in the colonial period established two fundamental principles: the concept of an official religion, that is, only one religion recognized and permitted by civil authorities; and the integration of Church and state. In the United States, of course, a fundamental principle of our political evolution was the *separation* of church and state. Moreover, many of the settlers who came to the English colonies came in search of religious freedom, not religious monopoly. As Samuel Ramos suggested,

> It was our [Mexico's] fate to be conquered by a Catholic theocracy which was struggling to isolate its people from the current of modern ideas that emanated from the Renaissance. Scarcely had the American colonies been organized when they were isolated against all possible heresy. Ports were closed and trade with all countries except Spain was disapproved. The only civilizing agent of the New World was the Catholic Church, which by virtue of its pedagogical monopoly shaped the American societies in a medieval pattern of life. Education, and the direction of social life as well, were placed in the hands of the Church, whose power was similar to that of a state within a state.[7]

The consequences of Mexico's religious heritage have been numerous. It is important to remember that Catholicism was not just a religion in the spiritual sense of the word but extended deeply into the political culture, given the influence of the Church over education and social organizations such as hospitals and charitable foundations, and its lack of religious competition.

One of the consequences is structural. In the first chapter, reference is made to the Mexican political system as corporatist. Corporatism extends back to the colonial period, when certain groups obtained special privileges from civil authorities, giving them preferred relationships with the state. Among these groups were clergy, military officers, and merchants. The most notable privileges received by the clergy were special legal *fueros,* or rights, allowing them to try their members in separate courts.[8] The Spanish established the precedent for favored treatment of specific groups. Once groups are thus singled out, they will fight very hard to retain their advantages. Much of nineteenth-century politics in Mexico became a battle between the Church and its conservative allies on just this score.

The monopoly of the Church in New Spain was very jealously protected. No immigrants professing other beliefs were allowed in before Mexican independence. The Church also took on another task for the state: ferreting out religious and political dissenters by establishing the Inquisition in the New World. The primary function of this institution was to identify and punish religious heretics, individuals who threatened religious beliefs as taught by Church authorities, but in practice the Inquisition controlled publishing, assembled a book index that censored intellectual ideas from abroad, and fielded special customs inspectors.[9] Although these activities were not entirely successful, in general the Church and civil authorities were intolerant of any other religious and secular thought. The Inquisition has been described in this fashion:

The belief that heretics were traitors and traitors were heretics led to the conviction that dissenters were social revolutionaries trying to subvert the political and religious stability of the community. These tenets were not later developments in the history of the Spanish Inquisition; they were inherent in the rationale of the institution from the fifteenth century onward and were apparent in the Holy Office's dealings with Jews, Protestants, and other heretics during the sixteenth century. The use of the Inquisition by the later eighteenth-century Bourbon kings of Spain as an instrument of regalism was not a departure from tradition. Particularly in the viceroyalty of New Spain during the late eighteenth century the Inquisition trials show how the Crown sought to promote political and religious orthodoxy.[10]

The heritage of intolerance plagued Mexico during much of its postindependence political history. It has been argued that because culturally there had been little experience with other points of view and in promoting respect for them, accommodation was not perceived as a desideratum. Some analysts suggest that the Catholic religion's continuation as a dominant presence in spite of religious freedom and the existence of other faiths encourages persistence of the phenomenon.

To carry out the conquest of New Spain, the Spanish relied on armed expeditions and missionaries. Once an area was made "safe" by an exploratory expedition, a permanent settlement around a mission and a *presidio,* or fort, was established. Some of the settlements were sited along a route known as the *camino real* (king's highway), which today is the old California Highway One. The original mission towns are now among the most important cities in the Southwest: San Francisco (Saint Francis), San Diego (Saint James), Santa Barbara (Saint Barbara), Albuquerque, Tucson, and Santa Fe.

Originally, the authorities used Spanish armed forces; in the colonial period, American-born Spaniards began filling officer ranks, as the government came to rely more heavily on the colonial militia. Although the armed forces were called on from time to time to protect the coast from French and British attacks, the army was primarily used to suppress Indian rebellions and to keep internal order. It patrolled the highways to keep them free of bandits. Basically, then, it functioned as police, not as defenders against external enemies.

The military, like the clergy, received special *fueros* in New Spain. It too had its own courts for civil and criminal cases, but unlike the clergy, military officers were immune to civil prosecution.[11] The favored status inevitably led to legal conflicts. Some historians have argued that one of the reasons for the disintegration of civil authority at the time of independence was declining respect caused by its inability to gain control over military cases.

As in the case of the Church, granting the military special privileges—which were passed on to the colonial militia before independence—created another powerful interest group. Their professional heirs in the nineteenth century wanted to retain the privileges. Furthermore, the close ties between military and civil authorities, and the unclear lines of subordination led to the blurring of distinctions in civil-military relations.[12]

In the nineteenth and early twentieth centuries, these patterns in civil-military relations and civil-Church relations impacted greatly on Mexico's political development. They complemented the corporatist heritage by establishing groups that saw their own interests, not those of society, as primary. The groups competed for political ascendancy, reinforcing the already-present social inequality by creating a hierarchy of interests and prestige.

To the legacies of corporatism, social inequality, special interests, and intolerance can be added the Spanish bureaucratic tradition. Critics tend to focus on the inefficiencies of the Spanish bureaucracy and the differences between the legal theory and the application of administrative criteria.[13] In part, problems can be attributed to the distance between the mother country and the colonies, as well as the distance between Mexico City, the seat of the Viceroyalty of New Spain, established in 1535, and its far-flung settlements in Yucatán, Chiapas, and what is today the southwestern United States. A more important feature of Spanish religious and civil structures was their strongly hierarchical nature and centralization. Low-level bureaucrats lacked authority. Decisions were made only at the top of the hierarchy. Delay, inefficiency, and corruption were the outcome.[14]

The hierarchical structure of the Spanish state in the New World is no better illustrated than through the viceroy himself. The viceroy (*virrey*) was in effect the vice-king, a personal appointee of and substitute for the king of Spain. He had two sources of power: he was the supreme civil authority and also the commander in chief. In addition, he was the vice-patron of the Catholic Church, responsible for the mission policies in the colonization process. Remember, this man, along with a second viceroy in Lima, Peru, governed all of Spanish-speaking Latin America and the southwestern United States.

The viceregal structure left Mexico with two political carryovers upon independence. First, the individual viceroys became extremely important,

---

Personalism: political authority and loyalty are given to an individual rather than to an institutionalized office held by a leader.

---

some serving for many years, completely at the whim of the crown. This shifted considerable political legitimacy away from Spanish institutions to a single individual. The personalization of power tended to devalue the institutionalization of political structures, enhancing the importance of political personalities. It also left Mexico with an integrated civil and religious/cultural tradition, complemented by an equally blended, hierarchical indigenous tradition of executive authority. Justo Sierra, a Mexican historian, described the viceroy's power, and the Church-state relationship:

> The Viceroy was the king. His business was to hold the land—that is, to conserve the king's dominion, New Spain, at all costs. The way to conserve it was to pacify it; hence the close collaboration with the Church. In view of the privileges granted by the Pope to the Spanish king in America, it could be said that the Church in America was under the Spanish king: this was called the Royal Patronate. But the ascendancy that the Church had acquired in Spanish America, because it consolidated, through conversions, the work of the Conquest, made it actually a partner in the government.[15]

Spanish political authority was top-heavy, placing most of the power in the hands of an executive institution. Few checks on the viceroy's decision-making authority existed. In many respects, the viceroy's self-developed political aura was equivalent to the *presidencialismo* described earlier. The Spanish did create an *audiencia,* a sort of quasi legislative-judicial body that acted as a board of appeals for grievances against the viceroy and could channel complaints directly to the crown, bypassing the viceroy. Also, the crown appointed its own inspectors, often secret, who traveled to New Spain to hear charges against a viceroy's abuse of authority. These *visitadors* were empowered to conduct thorough investigations and report to the crown.

The minor restrictions on viceregal powers did not mean there was separation of powers, an independent judiciary, a legislative body, or decentralization. Some participation at the local level existed, but Mexico had no legislative heritage comparable to that found in the colonial assemblies of the British colonies. Thus, it is not surprising that although Mexico quickly established a legislative body after independence, it functioned effectively for only brief periods in the 1860s and 1870s, and again in the 1920s, remaining ineffectual and subordinate throughout most of the present century.

Finally, another important Spanish political heritage is the role of the state in society. The strong authoritarian institutions in New Spain and the size of the Spanish colonial bureaucracy established the state as the preeminent institution.[16] The only other institution whose influence came close was the Catholic Church. Educated male Spaniards born in the New World essentially had three career choices: the colonial bureaucracy, the clergy (which appealed to only a minority), and the military. New Spain's private sector was weak, underdeveloped, and closed. The crown permitted little commercial activity among the colonies or with other countries. The monopolistic relationship between Spain and the colonies kept the latter from developing their full economic potential. Michael Meyer and William Sherman characterized Spain's policies as

> protectionist in the extreme, which meant that the economy in New Spain was very much restricted by limitations imposed by the imperial system. Thus the natural growth of industry and commerce was significantly impeded, because manufacturers and merchants in Spain were protected from the competition of those in the colony. In accord with the classic pattern, the Spanish Indies were to supply Spain with raw products, which could be made into finished goods in the mother country and sold back to the colonists at a profit. As a consequence, the character of the colonial economy in Mexico was essentially extractive.[17]

A long-term political consequence of a strong state and a weak private sector was the overarching prestige of the state, to the disadvantage of the private sector. Economically, then, the state was in the driver's seat, not because it controlled most economic resources but because it provided the most important positions available in the colonial world. The same mentality developed in the twentieth century in other colonial settings. For example, Indians came to believe that the British civil service was the preeminent institution in India, and government employment would grant them great prestige.[18] In the same way, positions in the Mexican state bureaucracy were seen by many educated Mexicans as the ultimate employment, and the competition for places was keen. One cultural theorist, Glen Dealy, argues that "public power like economic wealth is rooted in rational accumulation. Capitalism measures excellence in terms of accumulated wealth; *caudillaje* [Latin American culture] measures one's virtue in terms of accumulated public power."[19] This way of life did not end with the decline of the Spanish empire and Spain's departure from Mexico. Figures from the last third of the nineteenth century demonstrate that the government employed a large percentage of educated, professional men, suggesting again the limited opportunities in the private sector.

The Mexican state's importance can be explained not only by economic underdevelopment but by the status of the state in the New World. In other words, it was natural for Mexicans to expect the state to play an influential role. Not liking state intervention in their lives, similar to most people in the United States,[20] Mexicans nevertheless came to depend on the state as a problem solver, in part because the institutional infrastructure at the local level and the same self-reliant thinking was not present.

Spain bequeathed Mexico an individualistic cultural mind-set. North Americans, although characterized by self-initiative and independence, exhibited a strong sense of community. That is, throughout the western expansion, U.S. settlers saw surviving together as in the interest of the group as well as in the interest of its members. Mexicans, on the other hand, exhibited a strong sense of self. This, combined with sharper social-class divisions and social inequality, led to a preeminence of individual or familial preservation, unassociated with the protection of larger groups. The lack of communal ties reinforced the primacy of personal ties. It was a familiar phenomenon elsewhere in Latin America as well. In the political realm, it generally translates into *whom* you know rather than *what* you know. Although an almost universal truism, whom you know gains in importance where access to authority is limited.[21]

Finally, the structural arrangements of the Spanish colonial empire, the distances between the colonies and the mother country, and between the colonies themselves made for considerable dissatisfaction with the rules imposed. The Spanish settlers, and later their mestizo descendants, increasingly disobeyed the orders from overseas. Sometimes they could justifiably assert that a law no longer applied to the situation at hand. At other times they would flout a law they found inconvenient. The inefficiencies inherent in transatlantic management of possessions in two continents, built-in social inequalities, and the gap between Old World theory and New World reality meant the marginalization of Spain's laws in the Western Hemisphere. A lack of respect for the law and the primacy of personal and familial interests were fundamental factors in Mexico's political evolution from the 1830s through the end of the twentieth century.

NINETEENTH-CENTURY POLITICAL HERITAGE

Shortly after independence Mexico experimented briefly with a monarchical system, but the rapid demise of the three-hundred-year-old colonial structure left a political void. The only legitimate authority, the crown, and its colonial representative, the viceroy, disappeared. Intense political conflict ensued as various groups sought to legitimize their political philosophies. The battle for political supremacy affected the goals of the antagonists and influenced the process by which Mexicans settled political disputes. By the 1840s Mexico had fluctuated between a political model advocating federalism, the decentralization of power similar to that practiced in the United States, and centralism, the allocation of more decision-making authority to the national government.

As was true of many Latin American countries, Mexico was caught between the idea of rejecting its centralized, authoritarian Spanish heritage, and the idea of adopting the reformist U.S. model. The obstinacy of their proponents kept political affairs in constant flux. Violence was a frequent means for settling political disagreements, which enhanced the presence and importance of the army as an arbiter of political conflicts, and consumed much of the government budget that might otherwise have been spent more productively.

By the mid-nineteenth century two mainstreams of political thought confronted Mexicans: conservatism and liberalism. Mexican liberalism was a mixture of borrowed and native ideas that largely rejected Spanish authoritarianism and tradition, and instead drew on Enlightenment ideas from France, England, and the United States. Some of its elements included such basic U.S. tenets as guarantees of political liberty and the

---

Mexican liberalism: an amalgam of basic concepts of political liberty and nineteenth-century laissez-faire economic principles.

---

sovereignty of the general will. Among its principles were greater citizen participation in government, free speech guarantees, and a strong legislative branch. Liberals complemented these principles with a concept known as Jeffersonian agrarian democracy. Jefferson had advocated encouraging large numbers of small landholders in the United States. His rationale was that people with property constitute a stable citizenry; having something to lose, they would vigorously defend the democratic political process. The liberals also believed in classic economic liberalism, the philosophy pervading England and the United States during the same period. Economic liberalism of this period referred to the encouragement of individual initiative and the protection of individual property rights.[22]

Mexican conservatives held to an alternative set of political principles. Whereas an examination of Liberal ideas reveals most of them having been borrowed from leading thinkers and political systems foreign to Mexico's experience, the Conservatives praised the reform-minded Bourbon administration of the Spanish colonies prior to independence and emphasized a strong central executive. They argued for a strong executive because it would follow naturally after centuries of authoritarian colonial rule, and because the postindependence violence in the 1830s, 1840s, and 1850s seemed to be part of a larger struggle between anarchy and civilization in Latin America. Absent forceful leadership, Mexico would succumb to disorder and remain underdeveloped economically.[23]

The conservatives favored policies promoting industrialization, stressing light manufacturing rather than expansion of the small-landholder class. Mexico desperately needed capital, much of which had fled after independence and during the chaotic political period that followed. Both conservatives and liberals looked approvingly upon foreign investment and encouraged policies that would attract outside capital, particularly to mining and struggling industries such as textiles.[24]

Neither the conservatives nor the liberals gave much attention to the plight of the Indians. Because the thinkers in both camps generally were *criollos* of middle- and upper-middle-class background, their primary concerns were the maintenance of social order and the interests of their classes. Although the conservatives essentially ignored the Indians, the liberals sought to apply their philosophy of economic individualism to the Indian system of communal property holding, believing it to be an obstacle to development.

Liberals and conservatives clashed most violently on the role of the Catholic Church. The liberals believed, and correctly so, that the Church, as an integral ally of the Spanish state, conveyed support for the hierarchical, authoritarian, political structure.[25] Essentially, it was the Church's control of education and nearly all aspects of cultural life that permitted its influence. The conservatives, on the other hand, saw the Church as an important force, and worked toward an alliance with it.

Because the liberals viewed the Church as a staunch opponent and the conservatives' political and economic supporter, they wanted to reduce or eliminate altogether its influence. They introduced the Ley Lerdo (Lerdo law) on June 1, 1856, essentially forcing the Church to sell off its large landholdings, which at that time accounted for a sizeable portion of all Mexican real estate. The law did not have its intended consequences; the Church traded land for capital, thereby preserving a source of economic influence and at the same time enlarging the already substantial estates of the buyers.[26] The liberals also attacked the special privileges of the Church, which had been left inviolate by the 1824 Constitution immediately after independence. They eliminated its legal *fueros* and placed cemeteries under the jurisdiction of public authorities.[27]

From this brief overview, it can be seen that each side had something useful to offer. Yet their unwillingness to compromise and the intensity with which they held their opinions led to a polity in constant disarray. The battles between conservatives and liberals culminated in the War of the Reform, 1858–1861, in which the victorious liberals imposed their political views upon the defeated conservatives by force. These views are well represented in the Constitution of 1857, a landmark political document that influenced its revolutionary successor, the Constitution of 1917.

The issue of Church versus state, or the supremacy of state over Church, was a crucial element of the conservative-liberal battles, and a focus of nineteenth-century politics. The leading liberals of the day saw the classroom as the chief means of social transformation and the Church's control in that arena as undesirable; secular institutions were to be established. To implement this concept, President Benito Juárez appointed a committee under Gabino Barreda in 1867, an educator who established some basic principles for public education in the last third of the nineteenth century. Although the liberals hoped to replace Church-controlled schools with free, mandatory public education, their program was never fully carried out. Most important, they introduced a preparatory educational program, a sort of advanced high school to train future leaders in secular and liberal ideas.

Although the liberals succeeded in defeating the conservatives' forces by 1869, their unwillingness to compromise and their introduction of even more radical reforms—particularly those associated with suppressing the Catholic Church, and incorporated into the 1857 Constitution—impelled the conservatives and their Church allies to take the unusual step of seeking help from abroad. This ultimately led to the French intervention of 1862–1867, and an attempt to enthrone a foreign monarch, Maximilian. Although the liberals were nearly defeated during this interlude, under Benito Juárez's leadership they ultimately won out and executed Maximilian.

The liberals reigned from 1867 to 1876. This brief period is important because it provided Mexicans with a taste of a functioning liberal political model. The legislative branch of government exercised some actual power. Upon the death of Benito Juárez, his successors lacked the political skills and authority to sustain the government, and their experiment came to an end with the successful revolt of Porfirio Díaz, a leading military figure in the liberal battles against the French.

Díaz's ambition and his overthrow of Juárez's collaborators introduced a new generation of liberals to leadership positions. These men, most of whom were combat veterans of the liberal-conservative conflicts and the French intervention, were *moderate* liberals, distinct from the radical orthodox liberals of the Juárez generation. Díaz and the moderate liberals paved the way for the introduction of a new political philosophy into Mexico: positivism. As described by historian Charles Hale,

> [s]cientific or positive politics involved the argument that the country's problems should be approached and its policies formed scientifically. Its principal characteristics were an attack on doctrinaire [radical] liberalism, or

"metaphysical politics," an apology for strong government to counter endemic revolutions and anarchy, and a call for constitutional reform. It drew upon a current of European, particularly French, theories dating back to Henri de Saint-Simon and Auguste Comte in the 1820s, theories that under the name of positivism had become quite generalized in European thought by 1878. Apart from the theoretical origins of their doctrine, the exponents of scientific politics in Mexico found inspiration in the concrete experience of the contemporary conservative republics of France and Spain and in their leaders.[28]

The motto for many positivists in Mexico and elsewhere in Latin America was liberty and progress through peace and order. The key to Mexican positivism, as it was implemented by successive administrations under Profirio Díaz, who ruled Mexico from 1877 to 1880, 1884 to 1911, was order. After years of political instability, violence, and civil war, these men saw peace as a critical necessity for progress. Their explanation for the disruptive preceding decades centered on the nation that too much of Mexico's political thinking had been based on irrational or "unscientific" ideas influenced by the spiritual teachings of the Church, and that alternative political ideas were counterproductive.

Building on the philosophy of their orthodox liberal predecessors, the Díaz administrations came to believe that the most effective means for conveying rational positivist thought, or this new form of moderate liberalism, was public education. Education therefore became the essential instrument for homogenizing Mexican political values. It would turn out a new generation of political, intellectual, and economic leaders who would guide Mexico along the path of material progress and political development. Preeminent among the public institutions was the National Preparatory School in Mexico City, which enrolled children of regional and national notables. Its matriculation lists read like a roll of future national leaders.[29]

The acceptance of positivist ideas by the moderate liberals ultimately led to the dominance of order over liberty and progress. Indeed, it can be argued that after decades of civil conflict, positivism became a vehicle for reintroducing conservative ideas among Mexico's liberal leadership. Díaz increasingly used the state's power to maintain political order, allowing economic development to occur without government interference. His government encouraged the expansion of mining and made generous concessions to foreigners to obtain investment.

The *porfiriato,* as the period of Díaz's rule is known in Mexico, had significant consequences that led to the country's major social upheaval of the twentieth century, the Mexico Revolution of 1910, and numerous political and social legacies. Socially, Díaz attacked two important issues: the relationship between Church and state, and the role of Indians in the society.

Ironically, the Catholic Church regained considerable influence during the liberal era. Even Benito Juárez realized after Maximilian's defeat that pursuit of radical anti-Church policies would only generate further resistance and disorder. Díaz pursued a pragmatic policy of reconciliation in the 1870s, separating Church and state, but permitting Church to strengthen its religious role as long as it remained aloof from secular and political affairs.[30] Thus, the two parties achieved a *modus vivendi,* although the state remained in the stronger position, and the 1857 Constitution retained repressive, anti-Church provisions.

Díaz's attitude toward the Indians was also significant because it reflected a broader attitude toward social inequality. He and his collaborators, as did the original liberals, saw the Indians as obstacles to Mexican development. They applied the provisions of the law forcing the sale of Church property to the communal property held by Indian villages, accelerating the pace of sales begun by the orthodox liberals in the 1860s. But the positivists were not satisfied with this economic measure. Many of them accepted the notion, popular throughout Latin America at the time, that Indians were a cultural and social burden, and were racially inferior.[31] To overcome this racial barrier, they proposed introducing European immigration in the hope of wiping out the indigenous culture and providing a superior economic example for the mestizo farmer.

To ensure that immigration would take place, the Mexican government passed a series of colonization laws in the 1880s that granted generous concessions to foreigners who would survey public lands. By 1889 foreigners had surveyed almost eighty million acres, and had acquired large portions of the surveyed acreage at bargain-basement prices. For the most part, however, these individuals were not typical settlers; rather, they, like the Mexicans who purchased Church and Indian lands, were large landholders. Two million acres of communal Indian lands went to them and to corporations. Hence, the colonization laws not only increased the concentration of land in the hands of wealthy Mexicans and foreigners but antagonized small mestizo and Indian farmers, who became a force during the Mexican Revolution.

Although Díaz implemented policies that improved the country economically, the primary beneficiaries were the wealthy at home and abroad. The laboring classes, primarily mestizo in origin, benefited little from the politics of peace. Díaz focused on a small group of supporters and ignored the plight of most of his compatriots. Even middle-class mestizos, who rose to the top of the ladder politically by 1900, were limited in their abilities to share in the economic goods of the Díaz era.

For the purposes of understanding Mexican politics in the twentieth century, in the postrevolutionary era, it is even more important to explore the political heritage left by Díaz and his cronies. In the first place, although Church and state were separate, and the lines more firmly drawn between secular and religious activities, Díaz maintained fuzzier relationships between the state and two other important elements, the army and the private sector.

In effect, Díaz established the pattern for civil-military relations that characterized Mexico until the 1940s. Because he himself was a veteran of so many civil conflicts, it was only natural that he recruited many of his important collaborators, both on the national and state level, from among fellow officers.[32] Military men occupied many prominent positions. Although the presence of career officers in the top echelon declined across Díaz's tenure as they were replaced by younger civilian lawyers, no clear relationship of subordination between civil and military authorities was established. Díaz left a legacy of shared power and interlocking leadership.[33]

The unclear lines between military and civilian political power were duplicated between politicians and the business elite. Although it is the nature of a capitalist system to have an exchange of leaders between the economic and political spheres, as in the United States, such linkages in an authoritarian political structure, where access to power and decision making is closed, can produce potentially significant consequences. Díaz, who had control over most of the important national political offices, used appointments to reward supporters or as a means to co-opt opponents. At no time since 1884 has any administration had stronger elite economic representation in political office than under Díaz. Giving these positions, especially at the provincial level, to members of prominent families further closed paths of upward social mobility to less-favored groups, especially the mestizo middle class.

By the time Díaz began his third term as president in 1888, he had succeeded in controlling national elections, although he had not created a national electoral machine similar to that of the Partido Nacional Revolucionario (PNR) and its successors. He continued to hold elections to renew the loyalty of the people to his leadership, and to allow him to reward his faithful supporters with sinecures as federal deputies (congressmen) and senators. His control was so extensive that occasionally he chose the same person for more than one elective office.

Building on the original conservative philosophy and on the colonial heritage, Díaz reversed the tenuous decentralization trend begun under President Juárez. Structurally, he accomplished this by decreasing the powers of the legislative and judicial branches, making them subordinate to the executive branch, and to the presidency specifically. He also strengthened the presidency as distinct from the executive branch.

Díaz went beyond aggrandization of political authority in the executive branch and the presidency by strengthening the federal government or state generally. He did this by expanding the federal bureaucracy. Between 1876 and 1910 the government payroll grew some 900 percent. In 1876 only 16 percent of the middle class worked for the government; by 1910 the figure was 70 percent.[34] As in the colonial period, the private sector was not incorporating new generations of educated Mexicans; rather, their careers were being pursued within the political structure, notably the federal executive. Díaz provided the twentieth century with a dominant state, an apparatus most successful Mexicans would want to control because it was essential to their economic future.

Because Díaz held the presidency for some thirty years, a personality cult developed around his leadership. His collaborators conveyed the message that progress, as they defined it, was guaranteed by his presence. His indispensability enhanced his political maneuverability. On the other hand, Díaz put in place a political system that was underdeveloped institutionally. In concentrating on his personality, political institutions failed to acquire legitimacy. Even the stability of the political system itself was at stake because continuity was not guaranteed by the acceptability of its institutions but by an individual, Díaz.

The porfiriato also reinforced the paternalism handed down from the political and social culture of the precolonial and colonial periods. Díaz's concessions to favored individuals, providing them with substantial economic rewards, encouraged dependence on his personal largesse and the government generally. This technique, which he used generously to pacify opponents and reward friends, produced corruption at all levels of political life. It encouraged the belief that political office was a reward to be taken advantage of by the officeholder rather than a public responsibility. The political cultures of many other countries are similarly characterized to a greater or lesser degree.

Against his most recalcitrant foes, Díaz was willing to use less ingratiating techniques. Toward the end of his regime press censorship became widespread. Generally, he favored a controlled, complimentary press to counter criticism from independent sources. If threats or imprisonment were not sufficient to deter his opponents, he resorted to more severe measures. Typically, lower social groups were the victims of violent suppression. A notorious example of this policy was the treatment of the Yaqui Indians in northwestern Mexico, who rebelled after influential members of the Díaz administration began seizing their lands. The Yaquis were subjected to brutalities, were forced into what were in effect concentration camps, and many were deported to Yucatán, where most perished in forced labor on the henequen plantations in the hot tropical climate.[35]

As Mexico emerged from the first decade of the twentieth century, it acquired a political model that drew on Spanish authoritarian and paternal heritages. Like the viceroys before him but without reporting to any other authority, Díaz exercised extraordinary power. He built up a larger state apparatus as a means of retaining power, and although he strengthened the role of the state in society, he did not legitimize its institutions. While successfully building some economic infrastructure in Mexico, he failed to meet social needs and maltreated certain groups, thereby continuing and intensifying the social inequalities existing under his colonial predecessors. His favoritism toward foreigners caused resentment and contributed to the rise of nationalism after 1911. The lack of separation between civilian and military leadership left Mexicans unclear about the principle of civilian supremacy and autonomy, an issue that would confront his successors. Finally, it is important to emphasize that although the moderate liberals/converted positivists replaced orthodox liberals, and in many cases substituted conservative principles for their original political ideas, the excluded liberal followers who remained faithful to the cause rose up once again after 1910.

THE REVOLUTIONARY HERITAGE: SOCIAL VIOLENCE AND REFORM

It can never be forgotten that contemporary Mexico is the product of a violent revolution that lasted, on and off, from 1910 through 1920. The decimation of its population—more than a million people during the decade—alone would have left an indelible stamp on Mexican life. The Revolution touched all social classes, and although it did not impact on all locales with the same intensity, it brought together the residents of villages and cities to a degree never achieved before or since. In the same way that World War II altered life in the United States, the Revolution brought profound changes to Mexican society.

The causes of the Revolution have been thoroughly examined by historians. The causes are numerous, and their roots can be found in the

failures of the porfiriato. Among the most important to have been singled
out are foreign economic penetration, class struggle, land ownership, economic depression, local autonomy, the clash between modernity and tradition, the breakdown of the porfirian system, the weakness of the transition
process, the lack of opportunity for upward political and social mobility,
and the aging of the leadership. Historians do not agree on the primary
causes, nor do they agree on whether the 1910 Revolution was a "real"
revolution, that is, whether it radically changed the social structure.[36]

The Revolution, in my own view, introduced significant changes,
although it did not alter social structures to the degree one expects of a
major social revolution on par with the Soviet or Chinese revolutions.[37]
Nevertheless, for the purpose of understanding Mexican political developments in the twentieth century, it is essential to explore the ideology of the
Revolution, and the political structures that emerged in the immediate
postrevolutionary era.

Ideologically, one of the best ways to understand the diverse social
forces for change, is to trace briefly the constitutional provisions of 1917 to
the precursory and revolutionary figures. Among the most important precursors, Ricardo Flores Magón and his brothers offered ideas leading up to
the Revolution and revived the legitimacy of orthodox liberalism by establishing Liberal Clubs throughout Mexico.[38] This provided a basis for
middle-class participation in and support for revolutionary principles. Flores Magón and his adherents published a newspaper in exile in the United
States, *La Regeneración,* banned in Mexico. Many prominent political
figures in the Revolution, including General Alvaro Obregón, cited its
influence on their values. Perhaps more than in any other area, Flores
Magón provided arguments in support of workers' rights, establishing such
principles as minimum wage and maximum hours in strike documents and
Liberal Party platforms.[39] He also advocated the distribution of land, the
return of communal (*ejido*) properties to the Indians, and requiring agricultural land to be productive.

Politically, the most prominent figure in the pre- and revolutionary
eras was Francisco I. Madero, son of wealthy Coahuilan landowners in
northern Mexico, who believed in mild social reforms and the basic principles of political liberty. He founded the Anti-Reelectionist Party to oppose Porfirio Díaz. A product of his class, he did not believe in structural
change but did believe in equal opportunity for all.[40] His *Presidential
Succession of 1910,* the Anti-Reelectionist Party platform, and his revolutionary 1910 Plan of San Luis Potosí advocated three important political
items: no reelection; electoral reform (effective suffrage); and revision of
the Constitution of 1857. The most important of his social and economic

ideas concerned public education; he believed, as did the orthodox liberals, that education was the key to a modern Mexico.

More radical social ideas were offered by such revolutionaries as Pascual Orozco, who later turned against Madero, Francisco Villa, and Emiliano Zapata. Orozco, who expressed many popular social and economic ideas, some complementary to those of Flores Magón, also called for municipal autonomy from federal control in response to Díaz's centralization of political authority. Villa, from the northern state of Chihuahua, did not offer a true ideology or program, but the policies he implemented in the regions under his control reflected his radical social views. In Chihuahua, for example, he nationalized large landholders' properties outright, and because of his own illiteracy (he learned to read only late in life), instituted a widespread primary school program. Zapata, who came from the rugged state of Morelos immediately south of Mexico City, fought largely over the issue of land. His ideology, expressed by his collaborators, appeared in his famous Plan de Ayala.[41]

With the exception of Madero, these men offered few specific political principles. Consequently, the political ideology of the Revolution, with the possible exception of effective suffrage and no reelection, emerged piecemeal, either in the constitutional debates at Quarétaro, prior to the writing of the 1917 Constitution, or from actual experience.

One of the most important of these themes was Mexicanization, a broad form of nationalism. Simply stated, Mexico comes first, outsiders second. In the economic realm, it can be seen in stressing Mexicans instead of foreigners in management positions, even if the investment is foreign in origin. An even more important expression of economic nationalism occurred in regard to resources: the formalization of Mexican control. With few exceptions, at least 51 percent of any enterprise had to be in the hands of Mexicans. In 1988, desperate for foreign investment, the government loosened up some restrictions in certain economic sectors.

Mexicanization spread to cultural and psychological realms. On a cultural level, the Revolution gave birth to extraordinary productivity in art, music, and literature, in which methodology was often as important as

---

Mexicanization: a Revolution principle stressing the importance of Mexicans and Mexico, enhancing their influence and prestige.

---

the content. In the visual fields, the Mexicans revived the mural, an art form that could be viewed by large numbers of Mexicans rather than remain on the walls of private residences or inaccessible museums.[42] Political cartoons during and after the Revolution blossomed. In literature, the social protest novel, the novel of the Revolution, came to the fore. Often cynical, or highly critical, these works castigated not only the failures of the porfiriato but the apparent failures of the revolutionaries too.[43] Musicians paid attention to the indigenous heritage, even composing the classical *Indian Symphony,* whose roots lie within the native culture. Ballads and popular songs flourished throughout Mexico as each region made its contributions.[44]

Mexicanization also affected a line of intellectual thought known as *lo mexicano,* which was concerned with national or cultural identity, and pride in Mexican heritage. Henry Schmidt, one of the most insightful students of the Mexican cultural rebirth, assessed its impact:

> The 1910 Revolution generated an unprecedented expansion of knowledge in Mexico. At the same time as it lessened the tensions of an unresponsive political system, it ushered in a new age of creation. If the post-Revolutionary political development cannot always be viewed favorably, the efforts to reorient thought toward a greater awareness of national conditions at least merit commendation. Thus the 1920's is known as the period of "reconstruction" and "renaissance," when the country, having undergone its most profound dislocation since the Conquest, attempted to consolidate the gains its people had struggled for since the waning of the Porfiratio.[45]

Another important theme of the Revolution was social justice. Economically, although not expressed specifically in the Constitution, this included a fairer distribution of national income. Socially, and called for by nearly all revolutionary and intellectual thinkers, it involved expanded public education. Madero wanted to improve access. Many others promoted education as an indirect means to enhance economic opportunity, particularly for the Indians, whose integration into the mainstream mestizo culture could thereby be accomplished. A leading intellectual, José Vasconcelos, who made significant contributions to Mexican education, praised a coming "Cosmic race," suggesting that a racial mix would produce a superior, not inferior, culture.[46]

The Revolution did not react adversely to a strong state. Instead, building on the administrative infrastructure created under the porfiriato, postrevolutionary regimes contributed to its continued expansion. Yet, unlike Díaz, the Revolution heralded a larger state *role,* giving the state responsibilities unexpected of a government prior to 1910. For example, as a consequence of Mexicanization, the state gained control over subsoil resources and eventually became the administrator of extractive enterprises. The phenomenal growth in the value of the nation's oil in the 1970s cast the state in an even more important role. When the state nationalized foreign petroleum companies in 1938, it established national and international precedents elsewhere.[47] In later periods, the state came to control such industries as fertilizers, telephones, electricity, airlines, steel, and copper. In the mid-1980s the trend gradually began to be reversed.

The Revolution stimulated the political liberalism that had lain dormant under the ideology of positivism during the last twenty years of the porfiriato. Freedom of the press was revived during the Revolution. The media underwent a regression in the 1920s, and although censorship continued to raise its head, conditions under which they operated were much improved. The most important principle of political liberalism, increased participation in governance expressed through effective suffrage, was given substance in Madero's election in 1911, probably Mexico's freest, but has never returned to that level.

The political mythology of the Revolution, "Effective Suffrage, No Reelection," was stamped on official government documents until the 1970s. Effective suffrage is still only an ideal, however, not yet achieved in practice. On the other hand, no reelection, with but a few exceptions in the 1920s and 1930s, has become the rule. When General Alvaro Obregón tried to circumvent it in 1928 by forcing Congress to amend the Constitution to allow him to run again after a four-year hiatus, he was elected but then assassinated before taking office. No president since has tried the maneuver. No elected executive, including mayors and governors, repeats officeholding, consecutively or otherwise. Legislators may repeat terms, but not consecutively, a concept introduced in the 1930s.

Another revolutionary outcome was the changed relationship between Church and state. Once again, the seeds of orthodox liberalism appeared in the constitutional debates. Many of the revolutionaries eyed the Church with severe distrust and reinstituted many of the most restrictive provisions advocated by the early liberals. These provisions could be found, unchanged, until 1992 in the Constitution. They include removing religion from primary education (Article 3), taking away the Church's right to own real property (Article 27), and secularizing certain religious activities and restricting the clergy's potential political actions (Article 130). No clergy of any faith were permitted in their capacity as ministers to criticize Mexican laws or even to vote.

The breakup of large landholdings is also a primary economic and social product of revolutionary ideology. As part of the redistribution of land in Mexico after 1915, the government made the Indian *ejido* concept (village-owned lands) its own, distributing land to thousands of rural villages to be held in common for legal residents, who obtained use rights, not legal title, to it.[48] In effect, the government institutionalized the indigenous land system that the liberals and positivists had attempted to destroy. This structure remained unchanged until 1992. The Revolution also introduced a change in attitude toward labor. For the first time, strikes were legalized and the right to collective bargaining was sanctioned. Provisions regarding hours and wages, at least for organized labor, were introduced. The 1917 Constitution was the first to mention the concept of social security, although it was not implemented until 1943. Organized labor helped General Obregón defeat President Venustiano Carranza in the last armed confrontation of the revolutionary decade.

Finally, although this list is incomplete, the Revolution gave greater emphasis to a sense of constitutionalism. In a political sense, constitutionalism provides legitimacy for a set of ideas expressed formally in the national document. It is not only a reference point for the goals of Mexican society after 1920, as a consequence of the Revolution, but identifies the basic outline of political concepts and processes. The Constitution of 1917 itself took on a certain level of prestige. Although many of its more radical social, economic, and political provisions are observed more in abeyance than reality, its contents and its prestige together influenced the values of successive generations.[49]

THE POLITICS OF PLACE:
INTERFACE WITH THE UNITED STATES

The proximity of the United States has exercised an enormous influence on Mexico. As I will argue, "the United States constitutes a crucial variable in the very definition of Mexico's modern political culture."[50] Beginning with independence, the political leaders who sought solutions emphasizing federalism, and later the decentralizing principles of liberalism, borrowed many of their concepts from U.S. political thinkers and documents. In fact, the intellectual ideas provoked by U.S. independence from England provided a fertile literature from which independence precursors could also borrow.

The destiny of the two countries became intertwined politically in more direct ways as a consequence of the annexation of Texas, a northern province of New Spain. Immediately after Mexican independence, large numbers of Americans began to settle in Texas, quickly outnumbering the Mexicans there. The differences within Texas between Mexicans and Americans and between Texas and the Mexican government led to armed conflict. The Mexican army under General Antonio López de Santa Anna lay siege to the Alamo in February 1836 but was routed from Texas later that year. Texas remained independent of Mexico until 1845, when the United States, by a joint congressional resolution, annexed it. This provoked another conflict, one with even more serious repercussions.[51]

Desirous of more territory, President James Polk used several incidents as a pretext for war. In 1846 U.S. troops drove deep into Mexico's heartland, and in addition to occupying outlying regions of the former Spanish empire in New Mexico and California, seized the port of Veracruz and Mexico City. In the Treaty of Guadalupe Hidalgo, signed on February 2, 1848, Mexico ceded more than half its territory to the United States. Seven years later, the Mexican government, again under Santa Anna, sold the United States a strip of land (in what is now southern Arizona and southern New Mexico), known as the Gadsden Purchase, although this time not under duress.

The war left a justifiably bitter taste in the mouths of many Mexicans. As has been suggested, "the terms of the Treaty of Guadalupe Hidalgo are among the harshest imposed by a winner upon a loser in the history of the world."[52] More than any single issue, the terms established a relationship of distrust between the two nations. Physical incursion from the north took place twice more. Voices in the United States always seemed to call for annexations. Even as late as the first decade of the twentieth century, California legislators publicly advocated acquiring Baja California.

During the Revolution the United States repeatedly and directly or indirectly intervened in Mexican affairs. The intense personal prejudices or interests of its emissaries often determined U.S. foreign policy decisions. Henry Lane Wilson, ambassador during the Madero administration (1911– 1913), played a role in its overthrow, and in the failure to insure the safety of Madero and his vice-president, who were murdered by counterrevolutionaries led by Félix Díaz and Victoriano Huerta. Huerta established himself in power, and the violent phase of the Revolution began in earnest. President Woodrow Wilson removed the U.S. ambassador and sent personal emissaries to evaluate Huerta. He decided to channel funds to the Constitutionalists, revolutionaries who had remained loyal to Madero and to constitutional government. But after a minor incident involving U.S. sailors in the port of Tampico, Wilson used it as a pretext to order the occupation of the port of Veracruz, resulting in the deaths of numerous Mexicans.[53]

Wilson's highhandedness produced a widespread nationalistic response in Mexico that nearly brought Wilson's intention—to oust Huerta from the presidency—to naught. Mexicans alive at the time of the occupation recall discontinuing classes in English, switching back to Mexican cigarettes, and throwing away their Texas-style hats in symbolic protest. Young men as far away as Guadalajara, in western Mexico, readily joined voluntary companies to go fight the Americans.[54] But Huerta fell, and the North Americans did not invade and soon left Veracruz.

After the Constitutionalists' victory under Carranza, rebel chieftains began to bicker among themselves. They divided into two major camps: one led by Francisco Villa and Emiliano Zapata, the other under Álvaro Obregón and Carranza. After several major battles, Obregón defeated Villa's forces. In March 1916, after remnants of Villa's forces moved north and attacked Columbus, New Mexico, Wilson ordered a punitive expedition under General John "BlackJack" Pershing against Villa. The U.S. forces battled the Constitutionalists, never caught Villa, and remained in Mexico until 1917.[55]

From these necessarily brief, selected historical examples, it can be seen that Mexicans have reason to be distrustful of the United States. Even though Mexicans admire many qualities of their neighbor to the north, not least its political system, they have many reservations as well (see table 2-1).

Mexico's distrust of the United States because of the latter's involvement in Mexican affairs is complemented by cultural and economic relationships between the two countries. There is a lively commerce in ideas, music, art, fashion, and so on. The influence of U.S. culture can be found

Table 2-1 Mexicans' Perceptions of
the United States, 1989

What Mexicans Like Most	Percentage of Persons Interviewed
Economic opportunities	34
Cultural level	15
Democracy	14
Equality for all	8
Government protects the people	4
Its wealth	4
Good public services	3
Its liberty	2
Other reasons	1
Not sure	11
No answer	4

Source: Los Angeles Times poll, August 1989, courtesy of Miguel Basáñez. This poll is based on face-to-face interviews with 1,835 Mexican adults conducted from August 13th through 15, 1989, in 42 randomly selected towns and cities throughout Mexico. It has a margin of error of 3 percent in either direction.

worldwide, but it is rendered more intense in Mexico by virtue of closeness. Heavy tourist traffic runs in both directions. English is studied and spoken by many Mexicans. Spanish is the most popular foreign-language instruction in U.S. high schools, and Spanish classes from even the smallest rural communities often take field trips to Mexico. American music permeates the airwaves. American performers from rock stars to magicians are headliners in major Mexican cities. Television is saturated with shows and movies from north of the border, and the well-to-do who subscribe to satellite reception have access to a complete range of news and entertainment programs. These circumstances and others have contributed to a condition whereby

> the United States constitutes an almost unavoidable presence in the daily lives of most Mexicans. The music that is heard in the main urban centers of the country, the companies that dominate the billboards and television advertising, the entertainment and new broadcasts of the major media—all of these refer almost by necessity to the United States, creating a sense, however partial and distorted, of familiarity with U.S. society and culture.[56]

Throughout the twentieth century, especially since the 1930s, Mexicans have had to compete culturally and psychologically against American influence, which has had a serious impact on the structure of the intellectual community and on the content of university programs.[57] The proximity of place has also contributed to the fact that more Americans reside in Mexico than in any other Third World country, and more Mexicans reside in the United States than in any other country. Los Angeles has a higher concentration of people of Mexican descent than any city except Mexico City. It can be said that a greater percentage of Mexicans visited a First World country than did the citizens of any other Third World country. By 1989 one in three Mexicans had been to the United States, and nearly half said they had relatives living there.[58] Mexicans establish personal links to the United States at a level unmatched by any other country with the possible exception of Israel.

Culturally, the relationship is unequal. Because Spanish is the second-most-frequently spoken language in the United States, and Mexicans—along with Central Americans, Cubans, and other Latin Americans—have spread throughout the United States, their culture is influencing the culture of the United States. It can be found in music (for example, in the popular song "La Bamba," of which a hundred versions exist in Mexico), food, and art, and even language (*politico* comes to mind).[59] Some Americans have reacted with a heightened nationalism. The English First movement is an effort to reassert the supremacy of English and of traditional nonminority values in the face of what is deemed to be an onslaught of Hispanic values and language. Oscar Martínez writes,

> Groups such as U.S. English and English First have campaigned hard to convince a large portion of the American public that the use of languages other than English fosters fragmentation in the country and threatens future political stability. Led by California, by the mid-1980s about twelve states had declared English as their official language, sending a message to Hispanics and other language minority groups that they should rid themselves of their native tongues as quickly as possible. Thus far, similar measures in the legislatures of Texas, New Mexico, and Arizona have failed to pass, but the debate on the issue in these states is bound to increase in the future.[60]

The cheek-by-jowl closeness of Mexico and the United States has given rise to what many observers believe to be a distinct hybrid border culture. The contiguous regions, which share a range of serious problems from unemployment to pollution, have begun to develop local solutions together that reflect the hybrid culture. The borderland has been given the name "MexAmerica" by Lester Langley, who suggests that few in the United States are ready to admit that Mexico might be a determinant of the kind of society we are becoming and the character of our politics.[61]

Politically and culturally it is the United States that exercises the greatest long-term influence. Its position in the world gives it great prestige. This gives its political processes and to some extent its political values a certain degree of legitimacy in the eyes of many Mexicans, including politicians. Hence the implicit influence of the United States model, especially as the international trend toward political liberalization, that is, democratization, continues.[62] The influence also generates resentment. It has been argued convincingly by Jorge Castañeda that some Mexicans have felt uncomfortable supporting democratization because they sense it is a goal of the United States.[63]

Vexation with and suspicion of their neighbor engender a higher level of nationalism among Mexicans and among their leaders. Accordingly, some political decisions are driven by nationalism. Nationalism in its crudest sense is on the decline, as Mexicans' sense of national identity began gradually in the 1980s to extend beyond their country. Eighteen percent of all Mexicans conceptualize their sense of nationalism as extending beyond that of their own borders.[64]

One of the most sensitive issues in the U.S.-Mexican relationship, because of its prominence as a cause of the Mexican Revolution, is economic imperialism. Even as late as the early twentieth century, some U.S. diplomatic representatives had personal economic interests in Mexico, and these flavored their recommendations. By the beginning of the Revolution, U.S. investment in Mexico had reached staggering figures. Roger D. Hansen remarks that under Díaz foreign capital (38 percent from the United States) "flowed into the country in quantities proportionately much greater—in relation to national capital and the natural and human resources of Mexico—than the volume of European capital that entered the United States during its period of most intensive development. Only 100 million pesos as late as 1884, foreign investment rose to 3.4 billion by 1911."[65]

The ideology of Mexicanization, which grew out of the revolutionary struggle, traces its strongest roots to U.S. economic influence and the presence of American businessmen, managers, professionals, and land-owners, especially in the North. In formulating its economic development policies, Mexico, while protecting the sovereignty of its decision making, has had to consider its economic relationship with the United States. If the free trade agreement is completed, Mexico will be sharing its economic goals with those of the other two North American nations.

The impact of place is no better illustrated than by immigration. It is revealing because its consequences invoke cultural, economic, and politi-cal issues. Immigration has a long history and only in recent years has the United States attempted comprehensively and formally to prevent the flow of Mexican immigrants. Mexicans crossed the border to work for decades after the mid-nineteenth century.

Labor tends to follow capital, especially when unemployment and underemployment reach significant levels. Thus Mexicans went northward during difficult times in search of work. In the past some eventually be-came U.S. citizens, but most returned to their homeland after short stays. At other times when the United States faced labor shortages, especially of unskilled workers, it promoted controlled flows of such workers, which happened during World War II. What are the economic consequences of this relationship? For Mexico, it has meant increased economic oppor-tunities for its people, in some cases the potential for learning new skills. Also, their wages in large part have been remitted to their places of origin, thereby creating local sources of capital. For the United States, the eco-nomic benefits have been substantial. Historically, most of the migrants have worked in agriculture, providing cheaper foodstuffs than would have been otherwise available.

The political consequences of the relationship are numerous. For ex-ample, during periods of economic crisis in the 1980s in Mexico, millions of Mexicans sought employment in the United States, evidencing the in-ability of the political model to manage the economy. Indirectly, then, the United States aided Mexican domestic stability by channeling discontent northward. Yet Mexican nationalism, given the historical relationship discussed previously, forces Mexico to express concern over the migration. It is an embarrassment to Mexico. Some Mexicans have been maltreated in the United States or in crossing the border. Their exploitation pressures Mexico to place the issue on the agenda of Mexican-United States relations.

For the United States, the economic implications are obvious. Although the demand for unskilled workers now extends into a variety of occupational categories, critics charge that American workers go unemployed. Some groups in the United States resent what they see as their displacement by immigrant labor from Mexico. Moreover, employers are accused of hiring Mexican workers in order to avoid paying certain benefits and taxes.

Culturally, immigration spills over into many of the issues affecting U.S. Mexican relations. The English First movement came about largely through the rapid expansion of the U.S. Hispanic population, and the visible immigration of Mexicans and Central Americans into the Southwest. Immigration is a perceived cultural threat, which translates into local and state policy debates, often placed before voters. In Mexico, on the other hand, whole villages are literally ghost towns, as younger men and women leave for the United States. Children are often left in the hands of the mother alone or grandparents, breaking down the traditional family structure. Discouraged by the violence of U.S. cities, many migrant Mexicans who have children in the United States send them back to the home communities. These children bring American values with them, threatening the integrity of the local culture.[66]

Immigration has many more ramifications, both subtle and obvious. The point is that their proximity makes both countries prisoners, to some extent, of each other's problems. And although Mexico labors under a much greater dependency and subordination, it affects the United States. Mexico's political system must consider carefully the domestic issues and associated policies that bear on the relationship. The United States has no direct veto power over Mexican politics, but its presence casts a permanent shadow that the Mexican political leadership cannot ignore.

CONCLUSION

Throughout its recent history, Mexico, as a colony and independent nation, established patterns that have contributed heavily to the development of its political model. Some of the more important heritages from the Spanish colonial period have included the conflicts of social class, exacerbated by sharp social divisions. Catholicism, introduced as the official religion of the Spanish conquerors, was equally significant. Its monopoly encouraged a cultural intolerance toward other ideas or values, and enabled a symbiotic, profitable relationship between the state and the Church. The Spanish also fostered a strong sense of special interests, granting privileges to other selected groups, including the military, and ultimately contributing to a particularized civil-military relationship. These elements led to corporatism, a sort of quasi-official relationship between important occupational groups or institutions and the state. The Spanish, through their own political structure, especially the viceroy, imposed three hundred years of authoritarian, centralized administration. Great powers accrued to the executive, to the neglect of other government branches. Restrictive economic policies discouraged the growth of a strong colonial economy, thus shoring up the role of the state versus that of an incipient private sector. The state's power and prestige attracted New Spain's most ambitious citizens.

Many features of the colonial period were further enhanced after independence. The conflicts between the liberals and conservatives, driven by intolerance of counterviews, produced ongoing civil war and anarchy. Although Mexico experimented briefly with a more decentralized form of government, authoritarian qualities were back in the saddle by the end of the nineteenth century. The presidency replaced the viceroyship in wielding power, and President Díaz expanded the size and importance of the executive branch, thereby continuing to enhance the state's image. Although Díaz introduced political stability and some economic development, he perpetuated the social inequalities inherited from the Spanish period. He also insured the military a large voice in the political system, leaving the matter of military subordination to civilian authority unresolved. And the Spanish paternal traditions remained.

The Revolution reactively introduced changes but in many respects retained some of the basic features from the previous two periods. One important innovation was Mexicanization, an outgrowth largely of Mexico's exploitation by foreigners and especially its proximity to the United States. Mexicanization strengthened Mexican values and culture, as well as political nationalism. The Revolution altered the political rhetoric and social goals of Mexicans to legitimize the needs and interests of lower-income groups and Indians. Yet, instead of reducing the role of the state, it rendered the state an even more comprehensive institution. The Revolution also revived important principles of orthodox liberalism, including political liberties, suppression of the Church's secular role, and decentralization of authority, but a decade of civil violence and the need for effective leadership in the face of successive rebellions in the 1920s discouraged implementation of a federal, democratic system. Instead, the Revolution left Mexico with a heritage of strong, authoritarian leadership, of military supremacy. Even so, it established the importance of constitutionalism, even if many of the Constitution's liberal provisions went unenforced. The legitimacy of its concepts provided the basis for political liberalization under President Salinas.

Finally, Mexico's long, troublesome relationship with the United States has implications for its political evolution and the functioning of its model. The level of United States economic influence in Mexico, and the U.S. seizure of more than half of Mexico's national territory, prompted Mexican nationalism and anti-Americanism. Mexico has had to labor under the shadow of its internationally powerful neighbor, a psychological as well as practical political burden. Historical experience and geographic proximity flavored many domestic policy decisions, and perhaps subtly encouraged a strong, even authoritarian regime that could prevent the kind of instability and political squabbling that had left Mexico open to territorial depredation.

NOTES

1. Frank Tannenbam, *Mexico: The Struggle for Peace and Bread* (New York: Knopf, 1964), 36.

2. For an extensive discussion of racial relations in Mexico and elsewhere in Latin America, see Magnus Morner's classic study *Race Mixture in the History of Latin America* (Boston: Little, Brown, 1967).

3. Interestingly, this is even true when comparing the United States with its colonizer, England. See Richard Rose, *Politics in England* 5th edit. (Boston: Little, Brown, 1989), p. 69.

4. Charles Kadushin, *American Intellectual Elite* (Boston: Little, Brown, 1974), 26.

5. Judith Hellman, *Mexico in Crisis,* 2d ed. (New York: Holmes & Meier, 1983), 40–46.

6. For background, see Robert Ricard, *The Spiritual Conquest of Mexico* (Berkeley: University of California Press, 1966).

7. Samuel Ramos, *Profile of Man and Culture in Mexico* (Austin: University of Texas Press, 1962), 27.

8. Nancy Farris, *Crown and Clergy in Colonial Mexico, 1759–1821* (London: University of London Press, 1968).

9. For a fascinating account of the importance of imported books in the colonies, see Irving A. Leonard, *Books of the Brave* (Cambridge: Harvard University Press, 1949).

10. Richard Greenleaf, "Historiography of the Mexican Inquisition," in *Cultural Encounters, the Impact of the Inquisition in Spain and the New World,* ed. Mary Elizabeth Perry and Anne J. Cruz (Berkeley: University of California Press, 1991), 256–57.

11. Lyle McAlister, *The "Fuero Militar" in New Spain, 1764–1800* (Gainesville: University of Florida Press, 1967).

12. Edwin Lieuwen, *Mexican Militarism* (Albuquerque: University of New Mexico Press, 1968).

13. Henry Bamford Parkes, *A History of Mexico* (Boston: Houghton Mifflin, 1966), 87. For an excellent discussion of some of the consequences of the Spanish bureaucratic system, see Colin M. MacLachlan, *Spain's Empire in the New World* (Berkeley: University of California Press, 1991), 34ff.

14. For background, see Charles Gibson, *Spain in America* (New York: Harper, 1967); Clarence Haring, *The Spanish Empire in America* (New York: Oxford University Press, 1947); Lillian Fisher, *Viceregal Administration in the Spanish American Colonies* (Berkeley: University of California Press, 1926).

15. Justo Sierra, *The Political Evolution of the Mexican People* (Austin: University of Texas Press, 1969), 107.

16. *La formación del estado mexicano* (Mexico: Porrúa, 1984); Juan Felipe Leal, "El estado y el bloque en el poder en México," *Revista Mexicana de Ciencias Políticas y Sociales* 35 (October–December 1989): 12ff.

17. Michael Meyer and William Sherman, *The Course of Mexican History* (New York: Oxford University Press, 1991), 168.

18. Edward A. Shils, *The Intellectual between Tradition and Modernity: The Indian Situation* (The Hague: Mouton, 1961).

19. Glen Dealy, *The Public Man: An Interpretation of Latin American and Other Catholic Cultures* (Amherst: University of Massachusetts Press, 1977), 8.

20. In his recent examination of the heartland, William Least Heat-Moon reports that rural Kansas still strongly opposes any project representing federal government intervention. *PrairyErth* (New York: Houghton Mifflin, 1991).

21. For an excellent discussion of this in contemporary Mexico, see Larissa Lomnitz, "Horizontal and Vertical Relations and the Social Structure of Urban Mexico," *Latin American Research Review* 17, no. 2 (1982): 52.

22. For the views of a leading theoretician, and the larger context of liberalism in Mexico, see Charles A. Hale's *Mexican Liberalism in the Age of Mora, 1821–1853* (New Haven: Yale University Press, 1968).

23. For many interesting interpretations of the origins of authoritarianism, see John H. Coatsworth, "Los orígenes del autoritarismo moderno en México," *Foro Internacional* 16 (October–December 1975): 205–32.

24. For examples, see David M. Pletcher, *Rails, Mines, and Progress: Seven American Promoters in Mexico, 1867–1911* (Ithaca: Cornell University Press, 1958).

25. For long-term consequences of this relationship, see Karl Schmitt, "Church and State in Mexico: A Corporatist Relationship," *Americas* 40 (January 1984): 349–76.

26. Robert J. Knowlton, "Some Practical Effects of Clerical Opposition to the Mexican Reform," *Hispanic American Historical Review* 45 (1965): 246–56, provides concrete examples.

27. Jan Bazant, *Alienation of Church Wealth in Mexico: Social and Economic Aspects of the Liberal Revolution, 1856–1857* (Cambridge: Cambridge University Press, 1971).

28. Charles A. Hale, *The Transformation of Liberalism in Late Nineteenth Century Mexico* (Princeton: Princeton University Press, 1989), 27.

29. *Inscripciones,* Universidad Nacional Autónomo de Mexico, Escuela Nacional Preparatoria, official registration records.

30. Karl Schmitt, "The Díaz Conciliation Policy on State and Local Levels, 1867–1911," *Hispanic American Historical Review* 40 (1960): 513–32.

31. Martin S. Stabb, "Indigenism and Racism in Mexican Thought, 1857–1911," *Journal of Inter-American Studies and World Affairs* 1 (1959): 405–23.

32. For evidence of this, see the officer promotion lists from various battles in the published records of the Secretaría de Guerra y Marina, *Escalafón general de ejército* (Mexico City, 1902, 1911, 1914). For his collaborators, see my *Mexican Political Biographies, 1884–1934* (Austin: University of Texas Press, 1991).

33. For background, and the long-term consequences of this relationship, see my *Generals in the Palacio: The Military in Modern Mexico* (New York: Oxford University Press, 1992).

34. See Francisco Bulnes, *El verdadero Díaz y la Revolución* (Mexico: Editorial Hispano-Mexicana, 1920), 42. This latter figure is probably exaggerated, but indicates the bureaucracy's importance.

35. For a firsthand view of some of these methods, see John Kenneth Turner's muckraking, autobiographical account in *Barbarous Mexico* (Austin: University of Texas Press, 1969), or Evelyn Hu-Dehart, "Development and Rural Rebellion: Pacification of the Yaquis in the Late Porfiriato," *Hispanic American Historical Review* 54 (1974): 72–93.

36. An excellent but brief discussion of these arguments can be found in Paul J. Vanderwood, "Explaining the Mexican Revolution," in *The Revolutionary Process in Mexico: Essays on Political and Social Change, 1880–1940,* ed. Jaime E. Rodríguez (Los Angeles: UCLA Latin American Center, 1990), 97–114.

37. Support for this view can be found in John Womack, Jr., "The Mexican Revolution, 1910–1920," in *The Cambridge History of Latin America,* ed. Leslie Bethell vol. 5 (Cambridge: Cambridge University Press, 1986), 74–153.

38. For background on Flores Magón and other precursors, see James Cockcroft's excellent *Intellectual Precursors of the Mexican Revolution, 1900–1913* (Austin: University of Texas Press, 1968).

39. These can be found in Jesús Silva Herzog, *Breve historia de la revolución mexicana, los antecedentes y la etapa maderista* (Mexico: Fondo de Cultura Económica, 1960), annexes.

40. Stanley R. Ross, *Francisco I. Madero: Apostle of Mexican Democracy* (New York: Columbia University Press, 1955).

41. See John Womack, Jr., *Zapata and the Mexican Revolution* (New York: Knopf, 1968); Michael Meyer, *Mexican Rebel: Pascual Orozco and the Mexican Revolution, 1910–1915* (Lincoln: University of Nebraska Press, 1967).

42. Jean Charlot, *The Mexican Mural Renaissance, 1920–1925* (New Haven: Yale University Press, 1967), provides an overview of this movement. For its influence on United States culture, see Helen Delpar, *The Enormous Vogue of Things Mexican, Cultural Relations Between the United States and Mexico, 1920–1935* (Tuscaloosa: University of Alabama Press, 1992).

43. See John Brushwood's, *Mexico in Its Novel: A Nation's Search for Identity* (Austin: University of Texas Press, 1966), 173ff.

44. For wonderfully revealing examples of popular appraisals of various revolutionary figures, see Merle E. Simmons, *The Mexican Corrido as a Source for Interpretive Study of Modern Mexico, 1879–1950* (Bloomington: Indiana University Press, 1957).

45. Henry C. Schmidt, *The Roots of Lo Mexicano: Self and Society in Mexican Thought, 1900–1934* (Austin: University of Texas Press, 1978), 97.

46. José Vasconcelos, *La raza cósmica: misión de la raza iberoamericana* (Paris: Agencia Mundial de Librerías, 1925).

47. Paul Sigmund, *Multinationals in Latin America: The Politics of Nationalization* (Madison: University of Wisconsin Press, 1980), 81.

48. For an account of these developments, see Eyler N. Simpson's classic, *The Ejido: Mexico's Way Out* (Chapel Hill: University of North Carolina Press, 1937); Nathan Whetten, *Rural Mexico* (Chicago: University of Chicago Press, 1948), the most comprehensive picture of land-tenure conditions; Paul Lamartine Yates, *Mexico's Agricultural Dilemma* (Tucson: University of Arizona Press, 1981). In 1992, the Mexican government introduced radical reforms in the ejido land structure. See Claire Poole, "Land and Life," *Forbes,* April 29, 1991, 45–46, and *El Financiero International,* March 30, 1992, 11.

49. The best discussion of this consequence can be found in Frank Brandenburg, *The Making of Modern Mexico* (Englewood Cliffs, N.J.: Prentice-Hall, 1964), 10–11.

50. John H. Coatsworth and Carlos Rico, eds., *Images of Mexico in the United States* (La Jolla: UCSD Center for U.S.-Mexican Studies, 1989), 10.

51. For background, see Karl M. Schmitt, *Mexico and the United States, 1821–1973: Conflict and Coexistence* (New York: Wiley, 1974), 51ff.

52. Josefina Vázquez Zoraida and Lorenzo Meyer, *The United States and Mexico* (Chicago: University of Chicago Press, 1985), 49.

53. Robert E. Quirk, *An Affair of Honor: Woodrow Wilson and the Occupation of Veracruz* (New York: Norton, 1962), 95ff.

54. Interview with Ernesto Robles Levi, Mexico City, May 21, 1985.

55. For a firsthand account of this experience by a U.S. officer on the expedition, see Colonel Frank Tompkins, *Chasing Villa* (Harrisburg: Military Service Publishing Company, 1934).

56. Coatsworth and Rico, *Images of Mexico in the United States,* 10.

57. See my *Intellectuals and the State in Twentieth Century Mexico* (Austin: University of Texas Press, 1985), 79.

58. *New York Times* poll, October 28–November 4, 1986; *Los Angeles Times* poll, August 1989.

59. For numerous examples of these special influences, see Tom Miller, *On the Border: Portraits of America's Southwestern Frontier* (Tucson: University of Arizona Press, 1985).

60. Oscar Martínez, *Troublesome Border* (Tucson: University of Arizona Press, 1988), 97.

61. Lester D. Langley, *MexAmerica: Two Countries, One Future* (New York: Crown, 1988), 7.

62. For an excellent discussion of the dilemmas facing the United States government in the promotion of the liberal model in Mexico, see Sergio Aguayo, "Mexico in Transition and the United States: Old Perceptions, New Problems," in *Mexico and the United States: Managing the Relationship,* ed. Riordan Roett (Boulder, Colo.: Westview Press, 1988), 157.

63. Jorge G. Castañeda, "The Choices Facing Mexico," in *Mexico in Transition: Implications for U.S. Policy,* ed. Susan K. Purcell (New York: Council on Foreign Relations, 1988), 26.

64. Ronald Inglehart, Neil Nevitte, and Miguel Basáñez, *North American Convergence* (Princeton: Princeton University Press, forthcoming, 1993), chap. 6, 19.

65. Roger D. Hansen, *The Politics of Mexican Development* (Baltimore: Johns Hopkins University Press, 1971), 15–17.

66. A vast literature exists on this topic. Some of the best recent work, for example, is that of Jorge Bustamante, "Undocumented Immigration: Research Findings and Policy Options," in *Mexico and the United States,* ed. Riordan Roett (Boulder, Colo.: Westview Press, 1988), 109–32. For a comprehensive earlier account, see Wayne A. Cornelius, *Mexican Migration to the United States: Causes, Consequences, and U.S. Response* (Cambridge: MIT Center for International Studies, 1978).

3

Contemporary Political Culture: What Mexicans Value

What is problematic about the content of the emerging world culture is its political character. Although the movement toward technology and rationality of organization appears with great uniformity throughout the world, the direction of political change is less clear. But one aspect of this new world political culture is discernible: *it will be a political culture of participation* [italics added]. If there is a political revolution going on throughout the world, it is what might be called the participation explosion. In all the new nations of the world the belief that the ordinary man is politically relevant—that he ought to be an involved participant in the political system—is widespread. Large groups of people who have been outside of politics are demanding entrance into the political system. And the political elites are rare who do not profess commitment to this goal.

GABRIEL ALMOND AND SIDNEY VERBA, *The Civic Culture*

The political culture of any society is partially a product of its general culture. Culture incorporates all the influences—historical, religious, ethnic, political—that affect a society's values and attitudes. The political culture is a microcosm of the larger culture, focusing specifically on those values and attitudes having to do with an individual's *political* views and behavior.[1]

In the Mexican society, as in many societies, the intensity with which an individual holds certain values is related to religion, level of education, income, age, gender, place of residence, and other variables. Their impacts will be examined in the following chapter and are important to understand. Equally important for comparative purposes is to evaluate the beliefs that may influence Mexico's politics and Mexican attitudes toward the system.

LEGITIMACY: SUPPORT FOR THE
POLITICAL SYSTEM AND SOCIETY

One of the most significant explanatory variables regarding a political system's stability is its legitimacy in the eyes of the society. Of course, any political model consists of a variety of institutions, some of which having been accorded greater respect than others. Level of respect permits comparison of the standing of political and other types of institutions.

When Mexicans evaluate their institutions, it is apparent that those most closely associated with the state are held in lowest regard (see table 3-1). Only three institutions are widely esteemed: family, church, and schools.[2] The selection of family is not surprising because a culture with strong values generally ranks family and tradition highly. Of course, if loyalty to family is excessive, it makes transferring loyalty to governmental institutions difficult. This appears to be the case in Mexico, for Mexicans express some serious reservations about the trustworthiness of *governmental* institutions and institutions generally. The same pattern is found in Japan, where levels of trust in institutions is lower than in Mexico.[3]

The confidence Mexicans have in the Church and schools is significant. In the first place, as suggested in the previous chapter, the liberal and Revolution heritages both encouraged anti-Church sentiment. Nevertheless, although we will discover that Mexicans developed sentiments supportive of the separation of Church and state, secular criticism has not done away with respect or sympathies for the Catholic Church in particular in a society at least 85 percent of whose members are Catholic. Regard for the Church as an institution may be a partial reaction to state suppression. It may also follow from the Church's being one of the most autonomous institutions in the society, operating outside the control of the state despite severe constitutional restrictions. And it may well be that the Church largely earned its standing among Mexicans by its deeds. When Mexicans are asked to rank the most estimable individuals in their society, after parents, priests and schoolteachers are well above any others.

It is noteworthy that Americans give high marks to the church as an institution, indicating both their respect and, implicitly, the importance of religion and religious values in the U.S. culture. In England, on the other hand, where religion's influence is less controversial and less encompassing, churches are held in high esteem but closer to that found in Mexico.[4]

Mexican attitudes toward education, borne out in survey after survey, are generally quite positive. The significance of this for the legitimacy of

Table 3-1 Legitimacy of the State in England, United
States, and Mexico: Confidence of Citizens in Institutions

	Percentage Respondents Giving Positive Evaluation		
Institution	England	United States	Mexico
Family	—	—	84
Church	56	85	62
Schools	53	82	60
Television[a]	—	—	37
Law	—	—	32
Army	79	86	32
Newspaper/media	38	69	25
Businesses	55	84	22
Congress/parliament	52	83	16
Unions	29	52	14
Politics	—	—	12
Police	80	88	12

Sources: Este País, August 1991, 5; Laurence Parisot, "Attitudes about the
Media: A Five-Country Comparison," *Public Opinion* 10 (1988), table 1.
[a]For England and the United States, included under newspapers.

the political system is perhaps more important for Mexico than for the
United States and England, where schools are also viewed very positively,
especially in the former. The school system in Mexico is largely public,
although Catholic schools do play an important role. Unlike in the United
States, however, the public schools are operated by the national government and the teachers are its employees. Although they may not be perceived as doing so, they could serve as a positive, indirect means of
reinforcing the state's legitimacy—especially because texts in elementary
schools are selected by the government. Most important, Mexicans' satisfaction with the school system is one of the few consistent pluses for the
government.

Mexicans' confidence in other institutions is not prepossessing. What
one notices immediately comparing it with that of Americans is the generally lower levels of favorability. The weaker positive responses are not
necessarily an indication of extreme frustration with the Mexican system;
rather, Mexicans are likely to have lower expectations of their institutions,
given their institutions' past performances, than Americans have. Nevertheless, the fact that police, politics, and Congress tail off in the rankings
indicates a lack of confidence in as well as alienation from these institutions. In fact, in a 1987 study with additional categories, only 23 percent of
Mexicans gave government bureaucrats a favorable rating.[5]

Attitudes toward the police are an important indication of basic trust in government. On the local level, police are the most likely representatives of government to come in contact with the citizenry. Therefore, a good opinion of the police is generally seen as an important grass-roots indicator of trust in government. A sense of personal security is often a variable in one's evaluation of government performance. In both England and the United States the police achieved the highest level of confidence; in state and local surveys throughout Mexico, the police consistently rank lowest. Generally, explanations include the perception that they are dishonest, often involved in criminal activities, and abuse their authority, especially among lower-income and rural groups.

The connection between services and specific institutions in society is illustrated in table 3-2. When Mexicans are asked about the quality of specific government services, they point most often to education and health care. Generally speaking, they are most concerned on the local level, with education, health care, transportation, and sanitation. Agencies associated with the government render services that win widespread approval, thus contributing to the legitimacy of the government, and as in the case of the police and the security apparatus, render services that the average Mexican views as inadequate. The government connection is not the deciding variable in the evaluative process.

Mexicans' assessment of their most prominent institutions is unflattering as a rule, with the exception of Church and school. Unlike Americans, Mexicans do not highly regard private-sector institutions. This is in part due to the fact that private-sector values and the business community generally have not received positive attention in the schools or from public leaders. Indeed, businessmen are often denigrated. As one private-sector notable remarked, many Mexicans "use terms related to business, businessman, and entrepreneurs in a pejorative sense."[6] The ratings given to the private sector and to other institutions in 1991 are lower than in the mid-1980s, but the rank order is unchanged. The distrust is manifest.

Table 3-2 Legitimacy of the Mexican State:
The Case of Public Services, 1991

Service	Percentage Respondents with Favorable Image
Schools	67
Medical	55
Trash disposal and sanitation	41
Telephone	40
Security	32
Police	24

Source: Este País, August 1991, 4.

Table 3-3 Confidence in Government, Mexico,
United States, and Canada, 1981 and 1990

	Percentage Respondents Expressing Confidence in Government	
Country	1981	1990
Mexico	20	18
United States	50	36
Canada	38	34

Source: World Values Survey, 1990.

The attitude toward institutions may be explained in part by the grim economic and social conditions of Mexican life in the 1980s. A study in the early 1960s showed that urban Mexicans felt little pride in government institutions.[7] Yet, the 1980s were a decade of economic ups and downs, and confidence in government declined in many countries. In fact, if Canada, Mexico, and the United States are compared in this regard it is the United States that experienced a major decline (see table 3-3).

Although Mexicans' confidence in their government is half that of Americans and Canadians in their respective governments, its decline is minimal relative to the extent of economic crisis, and the economic conditions Mexicans faced during the 1980s. In 1986, 50 percent of Mexicans told interviewers they thought a revolution might occur by 1991. Nearly half also described their own economic situation in a mid-1980s *New York Times* poll as bad, and 11 percent as very bad. Nine in ten respondents believed the national economy was bad or very bad, and more than half thought it would not recover. Mexican attitudes toward government have remained remarkably stable. That confidence slipped no more than it did might be attributed to the high level of popularity achieved by President Salinas by 1990, reversing somewhat the decline in legitimacy of his predecessors' administrations. In other words, confidence in governmental institutions probably went below its 1990 level at some point between 1981 and 1990.

Mexicans expressed a much more favorable opinion of society in general than they did of specific institutions, governmental or otherwise, indicating a much higher level of trust in societal responses to problems. Scholars cite the 1985 earthquake in Mexico City as an example.[8] Although criticism abounds of government efforts to save persons trapped in the rubble, neighborhood volunteers' efforts are looked upon as exemplary. The same pattern was repeated in 1992 in the aftermath of a devastating explosion in Guadalajara's storm sewers.

The government's inadequacies after the quake eventuated in a groundswell of popular movements that together pressed demands on the government. One analyst had this to say about their cooperation:

> In the aftermath of the disastrous Mexico City earthquake in 1985, a coalition of urban organizations successfully forced the Mexican government and the World Bank to alter housing relief plans, accelerate the process of reconstruction, and reverse several fundamental urban policies. The coalition achieved this by uniting scores of neighborhood organizations. Hundreds of thousands of earthquake victims joined other urban poor to wrest concessions through deft media manipulation and political bartering.[9]

Another explanation for the government's faring poorly in the minds of most Mexicans is their perception of its goals. Asked in a survey in the late 1980s if government officials were working for their own interests or the interests of the majority, nearly two-thirds of the respondents said the former was the case. Cynicism characterized the replies.

Mexicans believe their society's qualities equal or superior to those of the U.S. society but are less certain concerning people. For example, in 1981 Mexicans had little confidence in their fellow human beings, about one-third that of Americans and Canadians (see table 3-4). When asked if one could trust the majority of people, fewer than one in five said yes.[10]

Interestingly, Mexicans' confidence in their fellow human beings almost doubled during the 1980s. (A slight increase also occurred among Americans and Canadians.) It is difficult to know to what this can be attributed. Dynamic social, economic, and political changes in recent years

Table 3-4 Confidence in Society and in People,
Mexico, United States, and Canada,
1981 and 1990

Country	1981	1990
	Percentage Respondents Expressing Confidence in Civil Society	
Mexico	47	48
United States	46	40
Canada	40	35
	Percentage Respondents Expressing Confidence in People	
Mexico	18	33
United States	45	50
Canada	49	52

Source: World Values Survey, 1990.

obviously influenced Mexicans' trust in different ways: trust in institutions, especially political institutions, declined; trust in society remained stable; and trust in individuals surged.

In terms of political behavior, trust in people is an important measure of the potential for democratic political institutions. Mexicans have expressed greater interest in democratizing their political institutions, sharing in the wave of democratization occurring elsewhere. To survive, democratic institutions rely on the high levels of personal trust necessary to effect compromise and operate within the rules of the political game. On a personal level Mexico has moved in that direction.

PARTICIPATION: ACTIVATING THE ELECTORATE

Trust in institutions and in fellow citizens is also related to political interest and participation. At least since the early 1960s interest in political affairs in urban Mexico has been lower than in the United States and England. Today, according to much better survey data, interest in politics remains relatively low. In 1986, 30 percent of all Mexicans expressed no interest in politics; 34 percent, little interest; and 36 percent, some or much interest. Differences between Mexico and the United States might be explained by differences in media, communications systems, and political competitiveness.

People generally move from an interest in politics to political activism when they believe they can affect outcomes in the system. One way to test peoples' attitudes toward outcomes is to examine *political efficacy*. This measures the degree to which a person believes he/she can participate in politics and the responsiveness of the system to their involvement. When Americans were asked whether or not they have a say in what government

---

Political efficacy: the belief in one's ability to participate in or influence political affairs.

---

does, somewhere between 33 and 41 percent, from 1980 to 1988, replied they do.[11] A similar but more specific question was posed to Mexicans; when asked if they could do something about election fraud, 56 percent thought not (see table 3-5). It is not surprising that more than half of all Mexicans believe they cannot affect the outcome of government policy; they live under a semiauthoritarian political model in which control over decision making is concentrated at the top. After all, if a third of all

Table 3-5 Political Efficacy of Mexicans, 1989

Response to Statement "Can do nothing about electoral fraud"	Percentage Respondents
Definitely true	8.9
True	47.3
False	31.5
Definitely false	4.0
Not sure	5.6
No answer	2.7

Source: Los Angeles Times poll, August 1989.

Americans described themselves as ineffectual politically in a system where honest elections are the norm and competition is regularized, the higher Mexican response should be expected.

Most citizens in political systems where elections occur become involved through voting. Therefore, their ability to affect the outcome of government policy is influenced by their perception of the integrity of the voting process. Mexico has had a long history of voter fraud in the twentieth century. Disputes over electoral results have occurred after every presidential election since 1920, and at the state and local levels as well. In 1929, 1940, and 1988 large numbers of Mexicans believed that the opposition candidate for president actually won the election.[12]

Before 1988 the accusations of wrongdoing were based solely on observation and political commentary,[13] but shortly before the presidential elections that year Mexicans were asked for the first time in a nationwide poll if their vote would be respected (see table 3-6). More than half of the interviewers thought their votes would not be counted honestly. Only a fourth believed in the integrity of the electoral process, and an equal number were unsure.

The question was repeated shortly before the off-year elections in August 1991, when many governors, half of the Senate, and all congresspersons were elected. Although the "don't knows" remained the

Table 3-6 Mexicans' Views of Elections, 1988 and 1991

Responses to Question "Will Your Vote Be Respected?"	Percentage Respondents		Percentage Change
	1988 Elections	1991 Elections	
Yes	23	42	83+
No	53	32	40−
Don't know	24	26	8+

Source: Este País, August 1991, 6.

same, those who viewed the elections as honest increased by 83 percent. Despite intense election battles and evidence of election fraud since 1989, the government successfully allayed the doubts of some Mexicans in the 1991 elections.

To test the assumptions about the relationship between citizens' perception of fraud and their willingness to vote, Mexicans were asked in 1989 their reasons for not voting. In Mexico, as in the United States, a person cannot vote unless he or she is registered before the election. Approximately half of all Mexicans eligible to vote were not registered. Although the government attempted to increase the number of registered voters by means of a voter identification card system instituted prior to the 1991 elections, many remained unregistered. The same situation obtains in the United States: in 1988, 33 percent of all persons eligible to vote were unregistered.[14]

The second-most-common reason Mexicans offer for not voting is that they forgot or were too busy (see table 3-7). If, using the sample in the *Los Angeles Times* August 1989 poll, we exclude the percentages of persons not voting because they were unregistered or forgot to vote and then collapse the percentages directly pertaining to the integrity of the electoral process (fraud, don't trust the process, and to protest election/candidate), the latter percentages would constitute 57.5 percent of the responses. It is also possible that a large percentage of those who never vote refrain because they believe that casting a ballot is meaningless.

People's proclivity to participate in the electoral process is affected to some extent not only by confidence in their political efficacy or by the integrity of the institutions and the process itself but by their level of activism in general. Mexicans' involvement in organizations is not high. Among the most important organizations are religious organizations and

Table 3-7 Mexicans' Reasons for Not Voting, 1989

Reasons for Not Voting	Percentage Respondents[a]
Not registered	45.7
Too busy or forgot	14.3
Fraud	7.4
Never vote	6.7
Don't trust process	5.4
To protest election/candidate	5.2
Too difficult	2.2
Too complicated	1.6
Don't want to be involved	1.4
Ill	1.4

Source: Los Angeles Times poll, August 1989.

unions, followed by charitable and youth groups. Fifty-eight percent of all Mexicans belong to no organization. With the exception of unions, membership is voluntary, which is a measure of level of interest in involvement. Slightly over half of all Mexicans who belong to organizations, belong to voluntary groups. Only 2 percent of Mexicans in the late 1980s belonged to political parties or political organizations, suggesting a relatively low level of interest in politics.[15]

The simplest form of political participation is voting, if it is one of the characteristics of the political model. A somewhat higher level of participation is indicated by membership in political organizations. Political organizations in Mexico have been constrained by the characteristics of the dominant government-party system and its control over the electoral process. In other words, most organizations expressing political goals or affiliated with a political party, especially at the national level, have ties to the political establishment. Two-thirds of all Mexicans do not belong to any political organization. Most of these individuals belong to unions and professional organizations that are automatically incorporated into the Institutional Revolutionary Party (PRI) (table 3-8). The most important of these are the party's sectorial organizations representing labor, peasants, and professional groups. Thus, working-class Mexicans are most likely to belong to the Mexican Federation of Labor (CTM) and the National Peasant Federation (CNC), and professionals and white-collar workers to the National Federation of Popular Organizations (CNOP). Others belong to such PRI organizations as youth or women's groups. Among Mexican activists who join political organizations, government employees account for 86 percent.

Given the fact that in Mexico government-controlled unions and professional organizations are part of the corporatist political structure, it is neither surprising that the government workers dominate political organization membership, and that membership is high. Few Americans are

Table 3-8 Membership in Political Organizations, Mexico, 1989

Organizational Affiliation	Percentage Respondents
Mexican Federation of Labor (CTM)	8.6
National Peasant Federation (CNC)	3.8
National Federation of Popular Organizations (CNOP)	4.5
Other PRI organizations	7.0
Independent (not government controlled)	4.0
None	66.7
Don't know	3.8
No answer	1.6

Source: Los Angeles Times poll, August 1989.

Table 3-9 Dispositions toward Political Action
in Mexico, United States, and Canada, 1981
and 1990

Country	1981	1990	Percentage Change
		Percentage Respondents Favoring Political Action[a]		
Mexico	7	24	242+
United States	15	26	73+
Canada	18	30	67+
		Percentage Respondents Actually Engaged in Political Action[a]		
Mexico	2	16	700+
United States	7	12	71+
Canada	9	16	78+

Source: World Values Survey, 1990.
[a]"Political action" includes boycotts, legal demonstrations, illegal
demonstrations, occupation of buildings or factories.

members of strictly political organizations. For example, a member of the
AFL-CIO (a major labor confederation) is not automatically a member of
either major party, although each party attempts to obtain labor's support
for its candidates and programs.

If we move higher up the ladder of political participation, from membership in an organization to some type of action, it is possible to obtain a
good sense of citizens' attitudes toward political involvement and of their
level of commitment to direct political participation. One way to measure
such participation is to ask citizens about their attitudes toward modes of
political action. In other words, to test receptivity to greater political
involvement, people are asked whether or not they favor such highly visible and committed activities as boycotting, legal demonstrations, illegal
demonstrations, and occupation of buildings or factories. Comparison of
attitudes in Mexico, the United States, and Canada appear in table 3-9.

As confidence in the institutions of government have declined in
Mexico, the United States, and elsewhere, the legitimacy of other forms of
political behavior has risen. People favor these actions because they believe
them to be effective means to convey political demands and, more important, that the regular channels are inadequate. In the past decade, in the
United States and Canada, a roughly 70 percent increase occurred among
those favoring less orthodox actions. Mexico experienced an even more
remarkable upsurge in that Mexicans traditionally have not favored such
activity. Among the many explanations of their quiescence is that their
more authoritarian system would not be likely to respond to the activity or

that it might respond repressively. Over time, the political culture came to see such activity as unacceptable and illegitimate. Whatever the explanation, only half as many Mexicans as Americans and Canadians favored such approaches in 1981, but the numbers approximated one another a decade later.

The dramatic increase among Mexicans accepting unorthodox political activity is indicative of the weakness of the Mexican system in coping with important demands, its decline in legitimacy, and perhaps its increased tolerance of such demands and its response to them. The percentage change in actual *engagement* in the unorthodox political activity is even more remarkable for the change it suggests in Mexican behavior. There has been a substantial increase of those willing to involve themselves directly in political activity in all three countries, even though the overall numbers remain small, fewer than one of six citizens. Still, Mexico's 700 percent gain is extraordinary. What explains this behavior and what does it mean?

The greater competitiveness of the national political game in Mexico beginning with local elections in 1985 and culminating in the national elections of 1988 had major repercussions on the nature of political activism. During the three-year period, public opposition to electoral fraud reached new highs.[16] It was given legitimacy in the media through announcements and advertisements by intellectuals and leading clergy. In fact, the clergy threatened after northern elections in 1986 to cease celebrating masses, something that had not happened since the 1920s, if a recount did not take place.[17] The claims of fraud attracted international attention, and the U.S. media helped to legitimize the claims of the domestic opposition. The rise of Mexican participation in unorthodox political activity to a level beyond that found in the United States suggests a major change in the political culture and in political efficacy. Although most Mexicans still think of themselves as ineffective politically, far greater numbers than before think they can induce change through non-government-controlled channels.

One of the most imaginative Mexican leaders of this new set of political techniques is "Superbarrio," a masked version of superman who is a wrestler. Sheldon Annis describes him as "a colorful good guy sworn to oppose the bureaucracy, greedy landlords, and political hacks. Dressed in yellow tights, red cape, and mask emblazoned with 'SB,' Superbarrio led tens of thousands of people in street protests over renters' rights, housing codes, construction credit, and low-cost housing."[18]

Most Mexicans, as do most Americans, however, participate politically through voting. Most Mexicans, on the other hand, do not support a

Table 3-10 Party Sympathy, Mexico, 1991

Party	Percentage Respondents
No party	56
Institutional Revolutionary Party (PRI)	28
National Action Party (PAN)	6
Democratic Revolutionary Party (PRD)	6
Other	4

Source: Este País, August 1991, 3.

political party. In fact, more than half of all Mexicans are what Americans label independent or uncommitted. In the United States in 1988, only 37 percent of all Americans considered themselves independent or uncommitted; 35 percent, Democrats; and 28 percent, Republicans. The higher percentage of Americans who are affiliated with a party is a consequence of a higher level of knowledge about the two major parties, which have operated during the entire century, and the fact that Americans have a choice.[19] Among Mexicans, 44 percent sympathize with specific parties (see table 3-10).

Most analysts describe the ideology of the three major parties as follows: PAN, right of center; PRI, center-right; and PRD, left of center. PRI would not always have received such a label. The breadth of its centrist posture allowed it to incorporate fairly radical populist views and presidents, such as Lázaro Cárdenas (1934–1940), as well as more conservative positions. Ideologically speaking, within their respective political systems, the United States and Mexico share certain similarities. Mexicans are more conservative politically in the 1990s, as is the United States electorate. But Mexicans are more strongly in the center ideologically, than are Americans. Whereas 45 percent of Americans described themselves as conservatives, 38 percent of Mexicans are sympathetic to the right. Mexico has a large group in the center, 44 percent; 31 percent of Americans placed themselves in this category. Liberals, who would not correspond precisely with Mexicans who lean to the left, account for 23 percent of the American electorate but only 18 percent of Mexicans (see table 3-11).[20]

Ideologically, people tend to select parties and other groups that reinforce their political views. For example, nearly half of all American conservatives are Republican, and half of all liberals are Democrats. Thirty-eight percent of all Mexicans are conservatives and nearly two-thirds identify with the PRI; more than half of those professing leftist views favor the PRD, the left-of-center party.

Although citizens tend to vote for political parties that they believe subscribe to their views, often ideology is not an important determinant of

Table 3-11 Relationship between Ideology and Party, Mexico and
the United States

Country and Ideology	Party Identification			Total
United States	Democratic	Independent	Republican	
Liberal	12	8	3	23
Center	11	13	7	31
Conservative	9	14	22	45
Mexico	Institutional Revolutionary Party (PRI)	Democratic Revolutionary Party (PRD)	National Action Party (PAN)	
Left	5	10	3	18
Center	25	7	12	44
Conservative	24	3	11	38

Sources: William Flanigan and Nancy Zingale, *Political Behavior of the American Electorate,* 17th ed. (Washington, D.C.: Congressional Quarterly Press, 1991), 107; World Values Survey, 1990, courtesy of Miguel Basáñez.

why people vote for a candidate. In fact, candidate ideology or program has very little to do with Mexicans' reasons for voting. Most Mexicans are interested in a particular candidate because they believe he or she will change things. The candidate's party is unimportant except for how the voter might perceive it as having a bearing on the candidate's ability to bring about change. In order of importance in supporting a candidate, Mexicans' reasons ranked as follows in 1989: to change things, 46.6 percent; ability, 14.7 percent; party, 8.4 percent; ideology, 8.0 percent; and other, 6.7 percent.[21]

POLITICAL MODERNIZATION:
AUTHORITARIANISM OR DEMOCRACY?

If ideology is not very important in how Mexicans vote, their values related to other issues play a critical role in the country's political development. Many Mexicans have long desired a more competitive political process, hoping to democratize the system. Desire, alone, of course, is not the only prerequisite of democracy. Democratization implies the importance of certain values. Some observers have remained skeptical because

> though there are few (and small) instances and spaces that are truly democratic in Mexico, these are exceptions to the overall trait of political life. The political culture, from its precedents in the Aztec world and through the

times of the Spanish colonial period, is characterized by values of subordination and authoritarianism. In the structure of the prototypical Mexican family, the father figure is authoritarian; children develop in an environment of domination that goes well beyond the natural figure of authority of parents over children in any family. These values are reproduced at the work place, at school, in the unions, in businesses, in political parties. In each and every realm of life, the individual members often criticize the authoritarianism of the government, even though they behave the same way in their own firms; unions complain about the hierarchical structures of firms and the government, though it would be hard to find a more hierarchical scheme of domination top-down than the corporatist structures of the labor sector. Most cases in which individuals carry out negotiations and transactions—which can be properly termed political—tend to be characterized by schemes of an authoritarian nature where there is always an implicit structure of domination.[22]

One of the most essential of these democratic values is a commitment on the part of over half of all Mexicans to expanding the role of opposition in Mexico (see table 3-12). Only one in four Mexicans believes the PRI should remain strong, about the same proportion who identify themselves as firm supporters of PRI. Other than PRI diehards, then, the typical Mexican is open to opposition-party growth.

Although the desire to see an expansion of opposition in Mexico is essential to political reform, an even more central feature of the actual *functioning* of a democratic political culture, as distinct from the institutions necessary to make it possible, is support for democratic liberties. Although not much research has been done on this variable in political liberalization, some comparative data are available (see table 3-13).

A 1978 survey of *urban* Mexicans and New Yorkers suggested that although Mexicans' beliefs in democratic values were not, in most cases, at the level of New Yorkers', strong support for most of the values existed.

Table 3-12 Attitudes toward Increased Political Opposition, Mexico, 1989

Answers to Question "Do you believe parties other than the governing party should increase their strength?"	Percentage Respondents
Yes	55.1
Only the Institutional Revolutionary Party (PRI) should remain strong	24.5
Not sure	13.2
No answer	7.2

Source: Los Angeles Times poll, August 1989.

Table 3-13 Political Authoritarianism: Support or Rejection of Democratic Liberties among Urban Mexicans and New Yorkers

Question and Response	Mean Score (Scale 1–10)[a]	
	Mexico	New York
Widespread Participation. Approve of		
Participating in petition signing	8.0	8.3
Participating in legal demonstration	8.2	8.1
Working for party, candidate, campaign	7.3	8.5
Support for Dissent. Approve of critics of Mexican government having the right to		
Vote	6.3	7.1
Hold peaceful demonstrations	7.6	7.0
Run for public office	4.7	5.7
Opposition to Suppression of Democratic Liberties. Approve of government's prohibiting critics of the Mexican political system from		
Holding public demonstrations	6.3	8.3
Holding meetings	6.4	8.1
Expressing views	6.4	7.8
Expressing views in media	6.4	8.3

Source: Adapted from John Booth and Mitchell Seligson, "The Political Culture of Authoritarianism in Mexico," *Latin American Research Review* 19, no. 1 (1984): 113, table 1.
[a] 1–5 = authoritarian; 6–10 = democratic.

The researchers examined three variables important to rejecting authoritarianism and supporting democracy: participation, political liberties, and dissent.[23] Not surprisingly, they found levels of support for participation, such as petition signing and demonstrating equal to those found among New Yorkers. The findings anticipated those discovered in the World Values Survey reported in table 3-9, in which Mexicans reached a level of support for such activities akin to that found in the United States. Regarding defense of political liberties, Mexicans scored in the democratic range, although not nearly as high as New Yorkers. In other words, Mexicans were not yet as tolerant of critics of the system. The most interesting results of this survey appear in connection with dissent.[24] The ultimate test of a democratic system is allowing a critic to run for public office. Neither New Yorkers nor Mexicans scored well on this measure.

When Americans were asked in another survey if everyone should have an equal right to hold public office, 91 percent replied yes.[25] But when they were asked in still another survey if Communists, with whom most Americans disagree intensely, should have an equal right to speak, only 64 percent said yes.[26] Although precisely the same questions have not been posed to Mexicans, one study asked if they would be bothered by someone's having different values and beliefs; 87 percent said no. But when asked if they would like a leftist living near them, the response indicated much less tolerance.[27]

In terms of political values, Mexicans share attitudes that are both conducive and resistant to democratization. Although some aspects of democratization may become integrated effectively into the culture, others are likely to be more difficult. Nevertheless, survey data do suggest a trend in Mexican values shifting strongly in a direction supportive of democratic behavior since the 1980s.

CONCLUSION

Values play a significant role in the evolution of a political system and the behavior of its citizens. Political values, as a component of general cultural values, are most important. In particular, three categories are central to the interrelationship between societal values and political behavior: legitimacy, participation, and authoritarianism.

Mexicans have high levels of respect for and trust in certain institutions, especially churches and schools, but Mexicans have very low levels of respect for political institutions of any sort and individuals associated with them, such as bureaucrats and police. Their appraisals reflect a general lack of trust in government. Mexicans are unusual, compared to Americans, for the low levels of respect they give to most societal institutions.

Although governmental institutions receive lower levels of support, and therefore have less legitimacy in Mexico than in the United States, the universal decline in governmental legitimacy in most nations during the 1980s was less sharp in Mexico. It is likely that Mexicans reached an even lower level of support for institutions in the mid-1980s, and through the efforts of President Salinas, who personally achieved high levels of popularity in the early 1990s, recovered from that level, thus considerably reducing the overall decline in government legitimacy.

Mexicans also expressed less trust than Americans or Canadians in their fellow human beings. Although their confidence in others rose dramatically during the 1980s, it is still substantially below that found in the United States. Nevertheless, Mexicans' remarkable increase in trust in one another is significant for their desire to increase participation and expand democratic institutions.

Other changes have also taken place in how Mexicans view their political efficacy. Although many are cynical about the election process and consequently their ability to influence government policy or leadership, a considerable shift occurred between 1988 and 1991 in the number of Mexicans who see the integrity of the process positively. Although more Mexicans view the election process as an accurate measure of their demands, the principal reasons for not participating, other than not being registered, relate to election fraud.

Not many Mexicans are highly active in voluntary social organizations. Even fewer are involved in political parties or organizations. Most of those who are politically involved are members of official party organizations—this is not surprising, given the politically interwoven corporatist structures discussed earlier. Of some surprise is Mexicans' increasing tolerance toward informal channels of political participation. Not only do they favor, to a much greater degree than in 1981, direct and unorthodox political actions on par with the level of support found in the United States but their actual involvement in such actions jumped 700 percent in the past decade.

The average Mexican, however, is not in favor of radical social and economic change but favors a peaceful, incremental approach. In fact, most Mexicans consider themselves moderate or conservative ideologically. Ideology itself is not an important determinant of party choices. Mexicans are more concerned about a candidate's willingness to improve conditions and his or her ability to do so.

Mexico is also part of a universal cultural shift described by Ronald Inglehart, Neil Nevitte, and Miguel Basáñez:

> [A] change from a world in which most people are absorbed in the tasks of sheer survival, to a world in which concern for the quality of life is becoming increasingly important. As we might expect, the peoples of Canada and the United States are well ahead of the Mexican public on this dimension, but during the 1980s all three publics showed substantial shifts toward increasing emphasis on Postmaterialist concerns.[28]

If Mexicans incorporate increased participation and the integrity of the political process in their definition of postmaterialist values, further interest in politics is likely to occur.

Finally, many Mexicans are interested in democratizing their political system, including increasing the competitiveness of the electoral process. Considerable evidence exists of support for increased political opposition. Less evidence is available on Mexican attitudes toward liberty, a crucial value in appraising the potential for successful democratization. Available data do suggest, however, the presence of many of the attitudes necessary for successful democratic behavior. They also suggest that Mexicans are still relatively intolerant of opposing views, and of allowing individuals with such views to participate politically and hold office.

NOTES

1. For a more comprehensive definition, see Walter A. Rosenbaum, *Political Culture* (New York: Praeger, 1975), 3–11.

2. In a comparative survey of Canada, Mexico, and the United States, Ronald Inglehart, Neil Nevitte, and Miguel Basáñez found that Mexico demonstrated greater levels of "strong" confidence in *nongovernmental* institutions. *North American Convergence* (Princeton: Princeton University Press, forthcoming 1993), figure 4-3.

3. Ibid., figure 4-2. Mexicans have also found it difficult to transfer loyalty to modern corporate forms in business, thus maintaining extensive family control. See Lavissa Lomnitz and Marisol Pérez-Lizaur, *A Mexican Elite Family, 1820–1980: Kinship, Class, and Culture* (Princeton: Princeton University Press, 1987).

4. Richard Rose, *Politics in England: Change and Persistence* (Boston: Little, Brown, 1989), 158.

5. Alberto Hernández Medina and Luis Narros Rodríquez, eds., *Cómo somos los mexicanos* (Mexico: CREA, 1987), 110.

6. Roderic A. Camp, *Entrepreneurs and Politics in Twentieth Century Mexico* (New York: Oxford University Press, 1989), 40.

7. Gabriel Almond and Sidney Verba, *The Civic Culture,* (Boston: Little, Brown, 1965), 64.

8. Carlos B. Gil, *Hope and Frustration: Interviews with Leaders of Mexico's Political Opposition* (Wilmington, Del.: Scholarly Resources, 1992), 48–57.

9. Sheldon Annis, "Giving Voice to the Poor," *Foreign Policy,* no. 84 (Fall 1991): 100.

10. Alberto Alvarez Gutiérrez, "Cómo se sienten los mexicanos?" in *Cómo somos los mexicanos,* ed. Alberto Hernández Medina and Luis Narro Rodríguez (Mexico: CREA, 1987), 81.

11. William H. Flanigan and Nancy H. Zingale, *Political Behavior of the American Electorate,* 7th ed. (Washington, D.C.: Congressional Quarterly, 1991), 180.

12. For discussion of each of these, see John Skirius, *José Vasconcelos y la cruzada de 1929* (Mexico: Siglo XXI, 1978), for 1929; Albert Michaels, "The Mexican Election of 1940," Special Studies No. 5, Council on International Studies (Buffalo: State University of New York, 1971), for 1940; Edgar Butler and Jorge Bustamante, eds., *Sucesión Presidencial: The Presidential Election of 1988* (Boulder, Colo.: Westview Press, 1991), for 1988.

13. For a brief discussion of polls and elections, see Miguel Basáñez, "Elections and Political Culture in Mexico," in *Mexican Politics in Transition,* ed. Judith Gentleman (Boulder, Colo.: Westview Press, 1987), 181–84.

14. Harold W. Stanley and Richard G. Niemi, *Vital Statistics on American Politics*, 3d ed. (Washington, D.C.: Congressional Quarterly Press, 1992), 88.

15. Alvarez Gutiérrez, "Cómo se sienten los mexicanos?" 87. Specifically, in order of response, Mexicans' organization memberships were religious groups, 17.6 percent; unions, 10.3 percent; charities, 7.8 percent; educational or artistic organizations, 4.1 percent; youth groups, 3.4 percent; professional associations, 2.9 percent; ecology organizations, 2.6 percent; parties or political groups, 1.9 percent; consumer advocacy groups, 1.7 percent; human rights organizations, 1.5 percent.

16. For background on this period, see Judith Gentleman, ed., *Mexican Politics in Transition* (Boulder, Colo.: Westview Press, 1987); Arturo Alvarado Mendoza, ed., *Electoral Patterns and Perspectives in Mexico* (La Jolla: U.S.-Mexican Studies Center, UCSD, 1987).

17. Javier Contreras Orozco, *Chihuahua, Trampa del Sistema* (Mexico: Edamex, 1987); Jaime Pérez Mendoza, "Por peteción de Bartlett, El Vaticano ordenó que hubiera misas en Chihuahua," *Proceso*, August 4, 1986, 6–13.

18. Annis, "Giving Voice to the Poor," 101.

19. Flanigan and Zingale, *Political Behavior of the American Electorate*, 52.

20. Ibid., 107; *New York Times* poll, 1986, courtesy of Miguel Basáñez.

21. *Los Angeles Times*, August 1989.

22. Luis Rubio, "Economic Reform and Political Liberalization," in *The Politics of Economic Liberalization in Mexico*, ed. Riordan Roett (Boulder, Colo.: Lynn Rienner, forthcoming 1993), 19–20.

23. For an extended discussion, see John Booth and Mitchell Seligson, "The Political Culture of Authoritarianism in Mexico," *Latin American Research Review* 19, no. 1 (1984), 106–24.

24. For comparisons with Costa Rica, a Latin American country that most Latin Americans consider "democratic," see Mitchell Seligson, "Political Culture and Democratization in Latin America," in *Latin America and Caribbean Contemporary Record*, ed. James Malloy and Eduardo A. Gamarra (New York: Holmes & Meier, 1990), A49–65.

25. Herbert McClosky and John Zaller, *The American Ethos: Public Attitudes toward Capitalism and Democracy* (Cambridge: Harvard University Press, 1984), 74.

26. Stanley and Niemi, *Vital Statistics on American Politics*, 28.

27. Alvarez Gutiérrez, "Cómo se sienten los mexicanos," 86.

28. Inglehart, Nevitte, and Basáñez, *North American Convergence*, chap. 7, 3.

4

Political Values and Their Origins: Partisanship, Alienation, and Tolerance

> Public support of the Mexican government is substantial. The few national interview studies that have been conducted show that the overwhelming majority of politically conscious Mexicans are positively allegiant to the nation, whatever criticisms and complaints they may have about specific institutions, practices, and men. However, it has also been argued—sometimes on the basis of the same data—that what the masses "give" to the system is not their support in any positive sense but rather their acquiescence, often expressed in noninvolvement and apathy. These two perspectives are not at all incompatible; both provide important, albeit partial, views of popular orientations to politics.
>
> RICHARD FAGEN AND WILLIAM TUOHY,
> *Politics and Privilege in a Mexican City*

Many experiences have bearing on the formation of values in general, and political values specifically. Values are general orientations toward basic aspects of life: abstract principles that guide behavior.[1] Children, for example, are affected by the attitudes of their parents, and most children carry

---

Values: general orientations toward basic aspects of life, abstract principles that guide behavior.

---

the consequences with them for years.[2] Other individuals have reported the influence of education and the specific role of teachers and professors.[3] Experiences other than those within the family and in school contribute to the formative years of many citizens, especially when the experiences are broad and deep, permeating the environment of an entire nation. The Great Depression, for example, tremendously affected Americans, their political and social values, and their voting behavior.[4] Undoubtedly, although we have no surveys to prove it empirically, the Revolution exerted a like influence in Mexico.[5] Some individuals, generally as young adults, consciously or unconsciously take on the values of their peers or of their working environment.

Although few studies exist of formative phenomena in Mexico, we know from studies of other countries that these are among the primary sources. We also know that certain variables most impinge upon political attitudes and values, and typically include race, ethnicity, socioeconomic background, level of education, occupation, region, and religion. Although Mexico has an Indian population, Indians account for only approximately 8 percent of the population, depending on the definition of *Indian*. Indians, however, are a minor political and economic presence, and hence have not been treated as a separate group in national political surveys. The typical Mexican thinks of himself as, and is, mestizo, thereby minimizing race, or ethnicity, as a significant variable in voting behavior. That circumstance could change if elections became more competitive and, more important, if Indians especially in certain states or regions were to organize themselves politically.

Because of sharp social-class divisions, Mexican values are likely to be affected by income level. Furthermore, the origins of Mexico's leaders, particularly political and economic, set them apart from the ordinary citizen. Consequently, it is important to ascertain differences between mass and elite political opinion. And because political knowledge has much to do with education, and disparities in schooling are substantial in Mexico, education is a way of distinguishing one Mexican from another and is strongly related to social class and occupation.[6] Historically, as suggested in chapter 2, regionalism played an important role in national politics. Although it had declined in prominence by the 1960s, it continues to exert an influence over some values, in the same way that it does in the United States. Religion is often still another determinant of political behavior, and in many societies plays a role in the formation of social and political values, especially when religious diversity is present. In Mexico, however, the predominance of Catholicism has obviated sharp religious differences. Most of the disharmony historically related to religion can be described as a battle between secularism and religion, not between religions. Nonetheless, the rise of Protestantism throughout Latin America since the 1960s, although not yet greatly felt in Mexico, and the presence of a small proportion of nonbelievers and atheists render religious beliefs deserving of consideration too.

INCOME AND POLITICS

The confidence people have in a political system and in their ability to influence the outcome of political decisions—level of political efficacy—depends on many things. One is income level. People who have achieved economic success not only perceive the system as fairer and more beneficial to their own interests but believe they can change aspects of it that they dislike. When Gabriel Almond and Sidney Verba published the first results from their multicountry study in the 1960s, they declared that Mexicans had a much lower sense of political efficacy than did Americans or the English but equivalent to that of the Germans.[7] In the late 1960s Rafael Segovia replicated research on political efficacy among schoolchildren and found that Mexican children were characterized by low levels of political efficacy. He also found that parental socioeconomic background had something to do with those levels; as parental income increased, so did the children's political efficacy and confidence in the system.[8]

Table 4-1 illustrates the 1989 levels of political efficacy nationally by income. Half the respondents with high incomes believed they could do something about electoral fraud. By contrast, only one in four low-income respondents so believed. In a study of residents of a state capital in the 1970s, Richard Fagen and William Tuohy also found very low levels of political efficacy; in fact, only 9 percent of the respondents thought they could do anything about a problem in their community.[9] Even so, the two researchers also discerned major differences based on income, with a sharp difference between upper-income and lower-income groups. The level of national political efficacy in the 1960s, when 38 percent of Mexicans thought they could change conditions, still holds in the 1990s, when 35

Table 4-1 Political Efficacy of Socioeconomic Status, Mexico, 1989

Response to the Statement "Can do nothing about electoral fraud"	Income Level			All Respondents (%)
	Low (%)	Medium (%)	High (%)	
Definitely true	11	6	7	8.9
True	50	47	32	47.3
False	27	38	47	31.5
Definitely false	3	5	8	4.0
Not sure	7	3	5	5.6
No answer	2	2	1	2.7

Source: Los Angeles Times poll, August 1989.

percent believe they can. In the United States the feeling of political efficacy is higher: in 1988, 54 percent believed they could affect government, a figure that has been more or less consistent since 1974.[10] Although income is more evenly distributed in the United States than in Mexico, lower-income Americans also indicate less political efficacy, but not as little as do similarly situated Mexicans.

These findings do not signify that Mexicans cannot overcome a low sense of political efficacy. As Ann Craig and Wayne Cornelius argue, low-income people who become active in nongovernmental organizations and make demands on the system, collectively develop a stronger sense of efficacy.[11] The growing numbers of such groups and their greater involvement, ultimately will increase participation and enhance political efficacy. Finally, economic growth itself, if it increases the proportion of Mexicans receiving higher incomes, likewise will enhance political efficacy.

The presence of some authoritarian values in Mexican culture was discussed earlier. In his survey of children in the 1960s Segovia found that authoritarianism was very much embedded in their value system. It is logical that individuals who have benefited least from the political system would turn toward development of opposition and alternative political choices. In fact, however, quite the reverse is true. Lower-income groups are slightly more *intolerant* toward opposition growth, and are least likely to encourage political alternatives. Regarding opposition parties, differences among income groups are slight, except when it comes to answering a question about whether they should be stronger (see table 4-2).

Analysts of Americans' voting behavior have always been attentive to variables affecting political sympathies for the Republicans and Democrats. Their studies suggest that among the most important is personal income. In the case of Mexicans, this was so in 1989 as well. The government party, the PRI, obtained its strongest support from upper-income voters; its weakest from low-income voters (see table 4-3). Although low-

Table 4-2 Political Tolerance by Socioeconomic Status, Mexico, 1989

| Response to Statement "Other parties should increase strength" | Income Level | | | |
	High (%)	Medium (%)	Low (%)	All Respondents (%)
Favor	62	62	54	56.3
Only the Institutional Revolutionary Party (PRI)	27	26	24	25.0
Not sure	7	10	16	13.5
No answer	4	2	6	5.2

Source: Los Angeles Times poll, August 1989.

Table 4-3 Partisan Sympathies by Socioeconomic Status, Mexico, 1989

| | Income Level | | | |
Sympathy for Party	High (%)	Middle (%)	Low (%)	All Respondents (%)
Institutional Revolutionary				
Party (PRI)	44	38	26	31.4
National Action Party (PAN)	21	13	12	13.1
Democratic Revolutionary				
Party (PRD)	5	16	17	15.5
Other	3	3	3	3.2
None	21	23	32	28.1
Don't know	4	5	7	6.4
No answer	1	1	2	2.2

Source: Los Angeles Times poll, August 1989.

income voters did not sympathize in large numbers with the populist, left-of-center opposition represented by the PRD, they proportionately represented its biggest constituency, although in percentages roughly equivalent to middle-income Mexicans. Sixteen percent of middle-income Mexicans sympathized with the PRD, compared with only 5 percent of higher-income Mexicans. By 1991, however, lower-income support for the PRD dropped to only 7 percent.[12] The PAN, considered by analysts to be a right-of-center party ideologically, not surprisingly attracts a disproportionate percentage of high-income sympathy. The August 1989 *Los Angeles Times* poll is also important for what it tells us about voter values. Individuals who receive higher incomes and who are more highly educated and who therefore are characterized by higher levels of political sophistication, are more decisive about their political choices and sympathies. They are the least likely to have no party sympathies. On the other hand, low-income Mexicans are more likely to be independent.

EDUCATION AND POLITICS

A variable closely related to income in determining political preference is education. Access to education, especially in a country like Mexico where opportunities are fewer than in the United States, is strongly related to parental income; the higher the income, the more likely a person will attend and *complete* higher education. For example, of the students at the National University in the early 1990s, over 90 percent were from families with incomes in the upper 15 percent.[13] Many Mexicans attend the public universities, the fees of which are minimal, but most low-income students

Table 4-4 Political Efficacy of Mexicans by Level of Education, 1989

Response to Statement "Can do nothing about electoral fraud"	Education Level				
	Primary (%)	Secondary (%)	Preparatory (%)	University (%)	All Respondents (%)
Definitely true	11	8	6	9	8.9
True	48	52	51	31	47.6
False	26	32	33	49	31.7
Definitely false	2	4	6	8	4.0
Not sure	10	3	3	2	5.6
No answer	4	1	1	1	2.2

Source: Los Angeles Times poll, August 1989.

do not complete the degree requirements. Students with higher education obtain the necessary credentials to pursue the most prestigious professions, just as they do elsewhere, and thus on the whole earn more.

With education come knowledge, social prestige, economic success, and greater self-confidence. Consequently, when Mexicans were asked whether they could do something about political fraud, that is, effect political change, nearly 60 percent of those with higher education believed they could (table 4-4). In contrast, a nearly equal percentage who had received primary education only believed they did not have the ability to change political conditions. Expressed differently, only half as many of those with primary education (28 percent) as with a college education or higher (57 percent), thought they could change political conditions. In the United States education affects responses in the same direction, and to the same degree.[14] Richard Fagen and William Tuohy, more than twenty years ago, discovered an even more exaggerated sense of political inefficacy among poorly educated Mexicans: only 9 percent thought they could change things.[15]

Education not only affects citizen confidence and knowledge about the political system but determines to some extent citizen acceptance of certain values. One of the values that education moderates is intolerance toward other political ideas and views. Nearly one in three Mexicans who have completed college education favors expansion of electoral competition (see table 4-5), thereby reducing the potential for authoritarian politics. More than any other variable, including income, level of education reduces political intolerance and lessens support for authoritarian political behavior. Educated Mexicans are sure where they stand on political issues, and they are committed to greater acceptance of nonstandard political views. More-available education in combination with higher incomes will chip away at the one-party-dominant system, which does not mean, how-

Table 4-5 Political Tolerance of Mexicans by Level of Education, 1989

Response to Statement "Other parties should increase strength"	Education Level				All Respondents (%)
	Primary (%)	Secondary (%)	Preparatory (%)	University (%)	
Favor	47	58	64	73	56.3
Only the Institutional Revolutionary Party (PRI)	28	24	23	18	25.0
Not sure	18	13	9	6	13.5
No answer	7	4	4	3	5.2

Source: Los Angeles Times poll, August 1989.

ever, that voter support for the PRI will concurrently automatically diminish.

Educational level affects party preference in Mexico as well, just as it has in the United States. In the latter education as a single variable does not have a dramatic effect,[16] but Mexico is characterized by sharper class divisions, making for a stronger relationship (see table 4-6). If both the PRI and the PAN are thought of as center-right alternatives, and the PRD as a center-left choice, the educational influence appears rather strong. Basically, as educational levels increase from primary to university, voters prefer the two traditional parties (the PRI and PAN). The inverse relationship is true for the PRD, which receives the least support from college-educated voters. However, the most exaggerated relationship is between college-educated voters and preference for the PAN, the party of the Right. Nearly twice as many college graduates as Mexicans in general prefer this party.

Table 4-6 Party Preference by Level of Education, Mexico, 1989

Preferred Party	Education Level				All Respondents (%)
	Primary (%)	Secondary (%)	Preparatory (%)	University (%)	
Institutional Revolutionary Party (PRI)	27	34	33	38	31.4
National Action Party (PAN)	10	13	14	24	13.1
Democratic Revolutionary Party	15	17	17	12	15.5
Other	3	4	3	2	3.2
None	35	24	24	18	28.1
Don't know	7	6	8	4	6.4
No answer	4	1	1	2	2.2

Source: Los Angeles Times poll, August 1989.

Again, as with the issue of political efficacy, Mexicans with low educational levels are the least likely to have a definite choice compared with only half as many college graduates who express no party preference.

RELIGION AND POLITICS

Students of the Catholic heritage in Mexico identify it as an important contributor to authoritarian values within the family and within the culture generally. When ranking the role of God in their lives, Mexicans and Americans give it equal importance; only one in four Canadians consider God important.[17] Religion's potential for influencing the formation of societal norms is enhanced by the fact that most Mexicans consider themselves religious (see table 4-7), and 85 percent declare they received a religious education in their homes.[18] Although it is true that the number of Mexicans who attend church services has declined since the turn of the century, the number who attend regularly is higher than typically is believed. In 1991, 44–45 percent of all Catholics went to church weekly or more often, and 14–19 percent monthly.[19]

Given the overwhelming dominance of Catholicism, it would be useful to measure its effect on political values according to the intensity of belief. For example, when Gabriel Almond and Sidney Verba completed their classic study, which largely ignored religion, they discovered that the more religious an individual, regardless of faith, the more intolerant of others' political beliefs.[20] Some observations will be offered on the variable of intensity, but for comparative purposes, it is helpful to identify the potential influence of religion on some of the major political values discussed above.

How does religion affect political efficacy? Table 4-8 presents responses according to religious belief. Because Catholics account for the

Table 4-7 Religious Affiliation, Mexico, 1989

Affiliation	Percentage of All Respondents
Catholic	92.2
Protestant	5.1
Other	.7
None	1.7
No answer	.3

Source: Los Angeles Times poll, August 1989.

Table 4-8 Political Efficacy by Religion, Mexico, 1989

Response to Statement "Can do nothing about fraud"	Catholic (%)	Religion Protestant (%)	None (%)	All Respondents (%)
Definitely true	9.1	2.6	12.8	8.9
True	47.4	46.8	46.2	47.3
False	31.3	35.1	30.8	31.5
Definitely false	4.1	0	5.1	4.0
Not sure	5.3	13.0	0	5.6
No answer	2.8	2.6	5.1	2.7

Source: Los Angeles Times poll, August 1989.

overwhelming majority of Mexicans, their views and that of the average Mexican are likely to correspond closely, but Protestant Mexicans, and those professing no religious beliefs, differ. First, Protestants' views of political efficacy are not nearly as definitive as those of Catholics; in other words, fewer thought they could definitely do something or nothing about fraud. They were more than twice as likely as Catholics to be unsure about their ability to affect political outcomes. Second, those professing no religion and those who were atheists had much stronger views. In fact, they were more likely to express views than religious Mexicans about the political process, and overall were more cynical about their ability to change conditions. Although no easy explanation exists for why Protestants are less sure of their political efficacy than Catholics are, the reason for stronger definitive responses among the nonreligious Mexicans can largely be attributed to education; the nonreligious are much better educated. Educated Mexicans are informed about the political process, and consequently tend to have more definite ideas. Education may also explain differences between Protestants and Catholics; Protestants as a group are less educated, which contributes to their greater indecisiveness.

The contribution of religion to authoritarian political values is embedded in the Mexican culture. It would be difficult, if not impossible, to separate religion from cultural values in general. The impact of different religions on political tolerance, however, can be measured. Among the three groups of Mexicans in 1989—Catholics, Protestants, and the nonreligious—Protestants supported the most liberal position, one favorable to the growth of opposition parties, although not to a significantly greater degree than Catholics (see table 4-9). Protestants' sympathies for the progressive political posture might be attributable in part to their self-perception as a fledgling minority in a dominant Catholic culture. Given their status in the religious realm, they may have a certain sympathy for minority political groups battling PRI's hegemony. Again, although Protestants

Table 4-9 Political Tolerance by Religion, Mexico, 1989

Response to Question "Do you believe parties other than the governing party should increase their strength?"	Religious Beliefs			All Respondents (%)
	Catholics (%)	Protestants (%)	None (%)	
Yes	55.1	59.7	43.6	55.1
Only the Institutional Revolutionary Party (PRI) should remain strong	25.1	15.6	20.5	24.5
Not sure	12.3	22.1	25.6	13.2
No answer	7.4	2.6	10.3	7.2

Source: Los Angeles Times poll, August 1989.

were less sure of their answers than Catholics, they gave very low support to the PRI as the only party. Growth in Protestantism would provide fertile grounds for opposition parties. The probability is especially significant because Protestantism's greatest inroads have been in lower-income, rural areas, the very districts where the PRI reports its heaviest support. Interestingly, the nonreligious also were less supportive of a single party than Catholics, but oddly, they did not favor a corresponding strengthening of other parties. A very high percentage were unsure, contradicting the unequivocal position of their answers on the issue of political efficacy and the rationale behind it.

What has most intrigued students of Mexican politics and religion is an assumed relationship between Catholicism and party affiliation. The reason for the assumption is that the National Action Party adopted many of the ideas of the European and Latin American Christian Democratic movements. Moreover, prominent early leaders of the party were known to be active Catholics.[21] Basically, the relationship between Catholicism and sympathy for the PAN is weak. In fact, as I pointed out in a more comprehensive examination of the issue, all of the survey data from the 1980s and early 1990s indicate that the only relationship between PAN and Catholicism is between the party and a tiny group of Catholics, 3.4 percent, who attend church daily. The group does differ from the rest of the population in intensity of support for the PAN, and higher proportions of votes cast for the 1988 PAN presidential candidate, Manuel Clouthier.[22]

Religious and party preferences are revealing, even if a tie between the PAN and Catholicism does not exist. Indeed, it is the nonreligious and the Protestants who exhibit reformist political sympathies, not Catholics. Using religion as a measurement, the strongest PAN supporters are Mexicans professing no religion (see table 4-10). The nonreligious Mexican is also the weakest supporter of the PRI, the establishment party. Mexicans

Table 4-10 Partisan Sympathies by Religion, Mexico, 1989

| Sympathy for Party | Religion | | | All Respondents |
	Catholic	Protestant	None	
None	27.5	35.1	28.2	27.9
Strong National Action Party (PAN)	5.2	5.2	7.7	5.2
National Action Party (PAN)	8.2	2.6	2.6	7.8
Strong Democratic Revolutionary Party (PRD)	6.6	10.4	12.8	7.0
Democratic Revolutionary Party (PRD)	8.4	9.1	7.7	8.5
Strong Institutional Revolutionary Party (PRI)	16.8	9.1	10.3	16.2
Institutional Revolutionary Party (PRI)	14.9	18.2	10.3	15.0
Other	3.2	1.3	5.1	3.2
No answer	8.9	9.1	15.4	9.1

Source: Los Angeles Times poll, August 1989.

professing no religion have given stronger-than-average support for the PRD, suggesting their sympathy for the newest, most radical opposition movement. Again, their higher levels of education make them more receptive to political alternatives.

Protestants illustrate diverse political sympathies, indicating they may be a more heterogeneous group in terms of background characteristics. Their support for the National Action Party is about half the support given by the average Mexican, regardless of religious belief. Protestants proportionately give their strongest preferences to the PRD, although not at the level of the nonreligious Mexicans. They too, therefore, are more sympathetic to the newest party and to the underdog. They give somewhat less support than average to the PRI but, more important, they rank much lower among those Mexicans *strongly* supportive of the PRI.

The data in table 4-10 suggest overall that non-Catholics, religious or otherwise, are more sympathetic to newer opposition parties, and that they are sources from which those parties can recruit successfully. The impact of religion on Mexican partisan politics, however, will continue to be moderated by the small numbers of non-Catholics. If Protestantism's growth were to mirror that found in Central America, where numbers have risen extraordinarily in the past two decades, religion could become a significant variable in Mexican voting behavior.[23]

Still on the subject of religion, it is important to note that religious issues have once again become prominent in the political arena at the behest of President Salinas. Such major issues as the right of priests to vote, legal recognition of the Church, and diplomatic relations with the

Vatican have been hotly debated in the 1990s.[24] Of all the issues involving the Church and the state in Mexico, the political role of the Church is the most controversial. Surveys from 1983 through 1990 make clear that anywhere between two-thirds to three-fourths of all Mexicans believe the Church should not participate in politics. Even among Catholics who regularly attend mass, Church political participation is firmly rejected. This is not to say that Catholicism, and the Church as an institution are not influential in Mexican life. Rather, Mexicans are a product of a liberal and Catholic heritage, and they distinguish between some liberal principles and other principles having to do with the Church and the role of religion (Catholicism) in their society.

GENDER AND POLITICS

One of the influences on values about which we have the least understanding is the role of gender in Mexico. A number of studies on Latin America have recently appeared that examine political behavior from a female viewpoint. Works on the political behavior of women in the United States have rarely discovered sharp differences with men, but they typically note that women are not as interested in politics, have less knowledge of politics, and are somewhat more alienated from the political system than are men. In fact, one study concludes that a high level of alienation was associated with the rise of feminism and recognition of their exclusionary treatment by the system.[25] Almond and Verba found the same pattern for Mexican women in the 1960s but with differences that were much more extreme.[26] For example, when asked if they discussed politics, 29 percent of Mexican women said yes, compared with 55 percent of Mexican men. In the United States, although fewer women than men discussed politics, the gap was relatively small: 70 versus 83 percent.[27]

Differences in political values and behavior attributable to gender can be explained by roles assigned to Mexican women. Although many women today obtain advanced education and a large percentage are in the work force, opportunities for women are fewer than for men. Moreover, the most detailed study of their values suggest that most women are as yet uncommitted to liberation and changing their traditional roles.[28] Given these and other conditions that have restricted women's roles in society and hence in politics, it is natural that they might feel more powerless to change the political system. A remarkable change in political efficacy seems to have occurred since the 1960s, however (see table 4-11). When Fagen and

Table 4-11 Political Efficacy by Gender, Mexico, 1989

Response to Statement "Can do nothing about electoral fraud"	Gender Male (%)	Female (%)	All Respondents (%)
Definitely true	9	9	8.9
True	47	49	47.6
False	31	32	31.7
Definitely false	6	2	4.0
Not sure	5	6	5.6
No answer	2	3	2.2

Source: Los Angeles Times poll, August 1989.

Tuohy carried out a study of Jalapa, Veracruz, in the 1970s, they found extreme differences between men and women regardless of social class. Typically, only half as many women as men reported high levels of political efficacy.[29] By 1989 almost no statistical difference existed between men and women on this issue. This finding corresponds with recent U.S. data on women and men.[30] Only in saying that the statement that they can do nothing about fraud is definitely false do Mexican women differ significantly from men.

As Mexicans make the transition from a more authoritarian political culture to one characterized by democratic characteristics, it is desirable to understand women's potential role. Given their smaller percentage of definitive responses, women might be expected to be more accepting than men are of new political alternatives. In fact, however, women differ little from men on the issue of political tolerance; indeed, they tend to be slightly more in favor of the status quo, and for a continuation of PRI dominance. Where women differ from men politically in Mexico is on political activism. Although few Mexicans have actually participated in some type of political protest, only half as many women as men have done so.[31]

Studies of European countries have generally found women to be somewhat ideologically more conservative than men. This does not hold for Mexican women today. However, more of them are uncommitted or support centrist views than men, and fewer identify with leftist political ideologies.[32] Typically, women everywhere are less interested than men in politics and hence participate less; this is true of Mexican women.[33] The ideological difference between men and women is translated into sympathy for political parties. Contrary to what some observers might allege, Mexican women express no more sympathy for the PAN, the conservative party, than do men. In fact, in a 1991 national survey, the PAN received stronger support from men than women. Women differ from men only slightly in support for the populist leftist party, the PRD, showing less sympathy.

REGION AND POLITICS

Many years ago Lesley Byrd Simpson wrote the classic *Many Mexicos*. *Many* in the title referred in large part to regionalism's influence on Mexican values. As Mexico developed and communications improved, regional differences declined, but they did not disappear. Economically speaking, the North is highly developed. It is characterized by heavy in-migration, dynamic change, industrialization, and of course, its proximity to and economic and cultural linkages with the United States. The South, on the other hand, is the least developed economically. It is rural, has a large Indian population, notably in Oaxaca and Chiapas, and is most isolated from the cultural mainstream. The center, which includes the Federal District, has been the traditional source of political leadership, religious infrastructure, industrialization, and intellectual activity.

Politically, regional differences have translated into electoral behavior. The National Action Party finds considerable strength in the North, where its more conservative economic platform has appeal. Baja California, located in this region, in 1989 elected the first PAN governor in this century. Many Mexicans believe that its industrialized sector, concentrated in the Nuevo León capital city of Monterrey, forms Mexicans who are products of a capitalist culture who hold attitudes different from those of Mexicans generally. The mythology is borne out by survey data. Repeated questions whose specific responses might connote stronger support for existing institutions suggest Northerners are much more likely to share that perspective than persons in other regions. For example, they have much more confidence in the legal system, they give much more positive marks to the armed forces, and they express a more favorable impression of police, who have the confidence of few Mexicans.[34]

In contrast, the South has been a bedrock of support for the PRI; indeed, without it the PRI would not have been able to sustain its victory, real or fraudulent, in the 1988 elections.[35] Because of the region's high percentage of agricultural workers with lower levels of income and education, one could expect that southerners would express lower levels of political efficacy than northerners. The data in table 4-12 show a sharp contrast between the two cohorts in terms of self-perceived ability to alter political conditions, especially if we consider only the true and false responses to the statement that one can do nothing about electoral fraud. About the same percentage from each region say they can do nothing. But northerners, more than any other cohort, believe they can do something.

Table 4-12 Political Efficacy by Region, Mexico, 1989

Response to Statement "Can do nothing about electoral fraud"	North (%)	Center (%)	Region Mexico City (%)	South (%)	All Respondents (%)
True	54	58	60	55	56.5
False	41	37	31	28	35.7
Not Sure	4	3	6	11	5.6
No answer	1	2	3	5	2.2

Source: Los Angeles Times poll, August 1989.

Nearly half again as many northerners as southerners are of this opinion. Southerners have the least confidence in their political effectiveness.

Place of residence can also affect other values, including religion, which in turn, as has been shown earlier, may have some effect on partisan political preferences. In terms of authoritarian values, the South, Center, and North are indistinguishable (see table 4-13). Only when respondents were asked about their sympathies for the major parties, did region produce important differences. In Mexico's most dynamic regions, those showing the highest levels of economic growth, the opposition, primarily the PAN, gained a stronghold. The PAN's primary source of sympathizers is the North and the Federal District, including the Mexico City metropolitan area. The South and the center with some exceptions, provide fewer sympathizers for this party. Contrary to what might be expected, the PRI obtains its greatest sympathy in the North and the Center. How does this square with election results, in which the PRI obtains most of the South's

Table 4-13 Partisan Sympathies by Region, Mexico, 1989

Sympathy for Party	North (%)	Central (%)	Region Mexico City (%)	South (%)	All Respondents (%)
Institutional Revolutionary Party (PRI)	41	35	24	18	31.4
National Action Party (PAN)	17	8	16	9	13.1
Democratic Revolutionary Party (PRD)	10	18	25	11	15.5
Other	5	2	4	1	3.2
None	21	30	19	48	28.1
Don't know	5	3	11	9	6.4
No answer	3	2	2	3	2.2

Source: Los Angeles Times poll, August 1989.

vote? The answer can be found among voters who sympathized with no party. Nearly half of all Mexicans residing in the South are independents without strong party sympathies; independents nationally account for a little more than one-quarter of all citizens. The PRD, at its apex in 1989, counted numerous adherents in Mexico City, and in several central states, notably Morelos and Michoacán. Again, residents of the North proved most decisive in their political views and were willing to express an opinion, followed by capital city residents. The data suggest that the South is a fertile region for opposition-party growth, if opposition parties can offer something the southerners seek.

AGE AND POLITICS

Age often determines important variations in values and, more important, suggests changes in the offing as generations reach political maturity.[36] In their significant comprehensive study, Inglehart, Nevitte, and Basáñez found that thirty-four issues had been characterized by intergenerational change in the past decade.[37] Inglehart and others discovered in earlier studies that economic conditions during a person's preadult years are the most significant determinant of adult values. Changing economic conditions, then, are likely to alter values from one generation to the next. For example, in the first World Values Survey, Inglehart learned that attitudes toward authoritarian values changed for each age cohort, moving in the direction of greater freedom and autonomy. The pattern peaked in all countries in the cohort aged twenty-five to thirty-four years old and began to reverse among the next generation.

Changing values in terms of obedience versus autonomy have been translated into political behavior, not only in Mexico but elsewhere in the Western world. In a previous chapter I remarked on the surge in unorthodox political activity in Mexico; conventional political participation more than doubled from 1981 to 1990.[38] The rise is attributable, in large part, to changing attitudes among *younger* age groups toward participation rather than among all age groups. This is particularly significant in a country where more than half the population is teenaged or younger.

Another consequence of generational change appears in party identification. As Inglehart reports, studies in Western Europe and in the United States demonstrate that older citizens identified more strongly with political parties, but that in recent decades younger voters are less likely to identify with a specific party. Although better educated than their elders and more interested in politics, younger Mexicans, like people elsewhere, no longer exhibit strong party loyalty. This phenomenon makes it difficult to predict future partisan sympathies and gives the independent or uncommitted voter considerable power to determine electoral outcomes, barring widespread fraud.

CONCLUSION

The foregoing brief analysis of just a few variables in the making of political values demonstrates the complexities of the research enterprise. Although many Mexicans have gained confidence in their ability to change the political system, large numbers believe themselves powerless. Those expressing the least confidence in their ability are women, the uneducated, and the poor.

Mexican values are undergoing change, and support for authoritarian structures is among those that are being recast. Younger people are contributing most to this alteration, as are those who are more highly educated, who come from affluent backgrounds, and who live in the most dynamic regions. Although many of these Mexicans are desirous of increasing political alternatives, including parties, it does not mean they necessarily would vote for any party other than the PRI.

The PRI continues to be viewed sympathetically by various segments of Mexican society. It has national strength, and by 1991 had recovered from a historic low in the 1988 elections. Its support continues especially solid in the South and to a lesser extent in the center. The appeal of the PAN was historically, and is in the 1990s, to people in the North, the Federal District, and selected states in the center. Mexicans are religious, but their Catholicism does not impinge upon their political behavior, their support of authoritarianism, or their partisanship. Many of the trends in political values as well as values in general that are apparent in Mexico are appearing in other countries as well, including the United States.

NOTES

1. Joseph A. Kahl, *The Measurement of Modernism: A Study of Values in Brazil and Mexico* (Austin: University of Texas Press, 1974), 8.

2. K. L. Tedin, "The Influence of Parents on the Political Attitudes of Adolescents," *American Political Science Review* 68 (December 1974): 1592.

3. Alex Edelstein, "Since Bennington: Evidence of Change in Student Political Behavior," in *Learning about Politics,* ed. Roberta Sigel (New York: Random House, 1970), 397.

4. Richard Centers, "Children of the New Deal: Social Stratification and Adolescent Attitudes," in *Class, Status and Power,* ed. Richard Bendix and Seymour Martin Lipset (New York: Free Press, 1953), 361.

5. For example, see such memoirs as Ramón Beteta, *Jarano* (Austin: University of Texas Press, 1970), and Andrés Iduarte, *Niño, Child of the Mexican Revolution* (New York: Praeger, 1971).

6. The most important variable determining the level of education a child obtains in Mexico is the socioeconomic status of the father, according to Kahl, *The Measurement of Modernism,* 71.

7. Gabriel Almond and Sidney Verba, *The Civic Culture: Political Attitudes and Democracy in Five Nations* (Boston: Little, Brown, 1965), 142.

8. Rafael Segovia, *La politización del niño mexicano* (Mexico: El Colegio de Mexico, 1975), 130.

9. Richard Fagen and William Tuohy, *Politics and Privilege in a Mexican City* (Stanford: Stanford University Press), 117.

10. Michael M. Gant and Norman R. Luttbeg, *American Electoral Behavior, 1952–1988* (Itasca: Peacock Publishers, 1991), 140.

11. Ann Craig and Wayne Cornelius, "Political Culture in Mexico, Continuities and Revisionist Interpretations," in *The Civic Culture Revisited,* ed. Gabriel Almond and Sidney Verba (Boston: Little, Brown, 1980), 369.

12. *Los Angeles Times* poll, September 1991, courtesy of Miguel Basáñez.

13. Ramon Eduardo Ruiz, *Triumphs and Tragedy: A History of the Mexican People* (New York: Norton, 1992), 469.

14. Gant and Luttbeg, *American Electoral Behavior,* 141, indicates that in 1988, only 25 percent of college-educated Americans reported little political efficacy, compared with 57 percent of those with less than a high school diploma.

15. Fagen and Tuohy, *Politics and Privilege in a Mexican City,* 117.

16. William Flanigan and Nancy Zingale, *Political Behavior of the American Electorate,* 7th ed. (Washington, D.C.: Congressional Quarterly Press, 1991), 68.

17. Ronald Inglehart, Neil Nevitte, and Miguel Basáñez, *North American Convergence* (Princeton: Princeton University Press, 1993), figure 3-20. Forty-six percent of citizens in the United States and 40 percent in Mexico consider God important in their lives.

18. *World Values Survey,* 1990.

19. Miguel Basáñez, *Encuesta nacional de opinión pública, iglesia y estado* (1990), 14.

20. Almond and Verba, *Civic Culture,* 101.

21. See Donald Mabry, *Mexico's Acción Nacional: A Catholic Alternative to Revolution* (Syracuse: Syracuse University Press, 1973), for the well-documented ideological influence. For the stereotypical allegation, without foundation, see Carlos Martínez Assad, "State Elections in Mexico," in *Electoral Patterns and Perspectives in Mexico,* ed. Arturo Alvarado (La Jolla: UCSD Mexico-United States Studies Center, 1987), 36.

22. See my "Clerics, Religion, and Political Modernization in Mexico" (Paper presented at the Annual Meeting of the American Political Science Association,

Washington, D.C., 1991). Charles Davis, in an analysis of survey data of Catholic workers in 1979–1980, made similar findings. See his "Religion and Partisan Loyalty: The Case of Catholic Workers in Mexico," *Western Political Quarterly* 45 (March 1992), 279.

23. For figures on this phenomenal growth, see David Stoll, *Is Latin America Turning Protestant? The Politics of Evangelical Growth* (Berkeley: University of California Press, 1990).

24. The president announced in his 1991 State of the Union address that new legislation governing the relationship would be forthcoming. *El Nacional,* November 2, 1991, 1. This legislation, reversing constitutional restrictions on the Church, was passed in early 1992.

25. Robert S. Gilmour and Robert B. Lamp, *Political Alienation in Contemporary America* (New York: St. Martin's Press, 1975), 55.

26. The best study using these data for Mexico is William J. Blough, "Political Attitudes of Mexican Women: Support for the Political System among a Newly Enfranchised Group," *Journal of Inter-American Studies and World Affairs* 14 (May 1972): 201–24.

27. Almond and Verba, *Civic Culture,* 327.

28. Enrique Alduncin, *Los Valores de los mexicanos, México: entre la tradición y la modernidad* (Mexico: Banamex, 1986), 189.

29. Fagen and Tuohy, *Politics and Privileges in a Mexican City,* 117.

30. Forty percent of men and 43 percent of women asserted a lack of political efficacy in 1988. Gant and Luttberg, *American Electoral Behavior,* 141.

31. *World Values Survey,* 1990, courtesy of Miguel Basáñez.

32. Ibid.

33. Ivan Zavala, "Valores políticos," in *Cómo somos los mexicanos,* ed. Alberto Hernández Medina and Luis Narro Rodríquez (Mexico: CREA, 1987), 97.

34. *World Values Survey,* 1990.

35. For an analysis of its importance, and regional support, see figure 5.3 and Joseph Klesner's discussion in "Changing Patterns of Electoral Participation and Official Party Support in Mexico," in *Mexican Politics in Transition,* ed. Judith Gentleman (Boulder, Colo.: Westview Press, 1987), 113ff. For the 1988 election results, see Edgar W. Butler et al., "An Examination of the Official Results of the 1988 Mexican Presidential Election," in *Sucesión Presidencial: The 1988 Mexican Presidential Election,* ed. Edgar W. Butler and Jorge A. Bustamante (Boulder, Colo.: Westview Press, 1991), 20.

36. Russell J. Dalton, *Citizen Politics in Western Democracies: Public Opinion and Political Parties in the United States, Great Britain, West Germany, and France* (Chatham, N.J: Chatham House, 1988), 85ff.

37. Inglehart, Nevitte, and Basáñez, *North American Convergence,* chap. 1, 12.

38. Ibid., chap. 4, 13.

5

Rising to the Top: The Recruitment of Political Leadership

> One of the most critical sets of questions about any political system concerns the composition of its leadership: Who governs? Who has access to power, and what are the social conditions of rule? Such issues have direct bearing on the representativeness of political leadership, a continuing concern of democratic theorists, and on the extent to which those in power emerge from the ranks of "the people"—or from an exclusive oligarchy. These themes also relate to the role of the political system within society at large, and to the ways in which careers in public life offer meaningful opportunities for vertical (usually upward) social mobility.
>
> PETER H. SMITH, *Labyrinths of Power*

Most citizens in a society where elections are typical participate by voting. A small number become involved in a political demonstration, or join a party or organization to influence public policy actively. An even smaller number seek political office and the power to make decisions.

The structure of a political system, the relationships between institutions and citizens, and the relationships among various political institutions affect how a person arrives at a leadership post. The collective process by which individuals reach such posts is known as political re-

---

Political recruitment: the collective process by which individuals reach political offices.

---

cruitment.[1] An examination of political recruitment from a comparative perspective is revealing for what it tells us about leadership characteristics and, equally important, what it illustrates about a society's political process.

All political systems and all organizations are governed by rules that prescribe acceptable behavior. The rules of political behavior are both formal and informal. The formal rules are set forth in law and in a constitution. The informal rules often explain more completely the realities of the process, or how the system functions in practice as distinct from theory. The political process melds the two sets of rules, and over time each influences the other to the extent that they often become inextricably intertwined.

THE FORMAL RULES

Formally, the Mexican political system has some of the same characteristics of the U.S. system. It is republican, having three branches of government—executive, legislative, judicial—and federal, allocating certain powers and responsibilities to state and local governments, and others to the national government. In practice, the Mexican system is dominated by the executive branch, which has not shared power with another branch since the 1920s, and allocates few powers to state and local governments, which in most places lack autonomy.

In a competitive, parliamentary system, such as that found in Britain, the legislative branch is the essential channel for a successful, national political career. The legislative branch is the seat of decision-making power and the most important institutional source of political recruitment. In the United States, decision-making power is divided among three branches of government, although in the legislative policy process both the executive branch and Congress play equally decisive roles. Not only is the structure in the United States different, measured by the actual exercise of political authority, but two parties have alternated in power.

The significance of these characteristics for recruitment is that they affect how candidates for office are chosen. The degree to which the average citizen participates effectively in the political process determines, to some extent, his or her voice in leadership selection. Of course, it is not just a choice between candidates representing one political organization or party versus another, but how specific individuals initially become candidates. The narrowness of the possible paths followed by potential political leaders in Mexico contrasts with the great breadth of approaches possible in the United States. The reason for the discrepancy is the dominance of a single political organization, the PRI and its antecedents, and a single leadership group within Mexico's political system.

In the formal structure, given the monopoly exercised by the PRI historically, one would expect the party itself to be crucial to the identification and recruitment of future political leaders. In actuality, its role has been minimal. The reason is that the PRI was not created nor has it functioned as an orthodox political party, that is, to control governance. The PRI, as suggested earlier, was formed to help *keep* a leadership group in power. Yet even a tight leadership group that exercises power in an authoritarian fashion must devise channels for political recruitment. Not to do so would eventually deprive it of the fresh replacements necessary to its continued existence.

When individuals in a small group exercise power over a long period of time, they tend to develop their own criteria for selecting their successors.[2] Moreover, they personally exercise the greatest influence over the selection process. Students of political recruitment call this *incumbent, or sponsored, selection.*[3] In other words, other groups in the society, such as voters, do not really have a decisive voice concerning future leaders.[4]

Sponsored selection: political recruitment dominated by incumbent officeholders.

Sponsored selection is well illustrated by the Mexican candidate-selection process for the presidency. Formally, presidential candidates of the government party (the PRI) are chosen by party delegates in what appears to be an internal, democratic procedure. In reality, party rank and file, and even party leadership have little, if anything, to do with it. The incumbent president actually designates his successor, who then becomes the PRI nominee. Although characteristics of presidential selection varied slightly in the past decade, the essential qualities persist.[5] Basically, one way or another, in the last third of an administration, the names of possible nominees emerge. They are always members of the cabinet, implying of course, that membership in it is essential to the career of a supremely ambitious politician. In contrast, rarely has a U.S. president held a cabinet post. For example, Jimmy Carter won the Democratic nomination after serving as a state governor. Ronald Reagan had been a governor. Neither had ever held a national appointive political office. George Bush's career most closely matches a Mexican president's political career—largely appointive, and in the executive branch. Even Bush, however, reached the presidency from *elective,* not appointive, office.

Once several cabinet members are openly talked about as potential presidential candidates, politicians within the government and PRI supporters identify with the one who they hope will become the incumbent president's choice. Under President Miguel de la Madrid (1982–1988), each of several contenders gave a formal presentation to Congress, and each organized separate conferences to express their policy views publicly. The strongest potential candidates were Manuel Bartlett, secretary of government; Alfredo del Mazo, secretary of energy; and Carlos Salinas, secretary of programming and budgeting. De la Madrid chose Salinas as the PRI candidate, amidst considerable controversy within the political leadership, and Salinas organized his own campaign.

The incumbent president's designation of his successor not only places extraordinary power in the hands of the chief executive but reinforces the centralization of political power in general. The executive privilege also flavors the nature of incumbent recruitment at all levels of the political system.

THE INFORMAL RULES: WHAT IS NECESSARY TO RISE TO THE TOP

If neither the party nor the electorate is significant in Mexican political recruitment, which institutions are? Even when individuals determine the outcome of the process, from the president on down, institutions facilitate the initial stages of a prospective leader's career. Strangely, the most important institution in the initial recruitment of political leaders is the university.

Mexico's postrevolutionary leadership, building on the concept of the National Preparatory School introduced by the liberals in the mid-nineteenth century, used public education as a means of preparing and identifying future politicians. Many of the prerevolutionary leaders had been educated at the National Preparatory School in the nineteenth century, and it continued so to function after 1920. Some politicians who served in national posts in the 1920s and 1930s never obtained higher education; they were self-made, largely on revolutionary battlefields from 1910 to 1920. Many continued as career military officers in the new postrevolutionary army.

A rapid shift occurred in credentials between the revolutionary generation of political leaders (holding office from 1920 to 1946) and the postrevolutionary generation (holding office from 1946 through the 1960s). The importance of higher education in political recruitment and the rapid decline of battlefield experiences are clearly illustrated by the personal experiences of presidents Lázaro Cárdenas (1934–1940) and Miguel Alemán (1946–1952) and the individuals they recruited to political office. Cárdenas joined the Revolution as a young man and rose through the ranks to become a division general, Mexico's highest-ranking officer. He had no formal education beyond primary school in his hometown. Although pursuing a political career in the 1920s, he remained in the army, eventually serving as secretary of defense. Miguel Alemán, son of a prominent general, on the other hand, was too young to have fought in the Revolution. Encouraged by his father to get a good education, he was sent to Mexico City where he studied at the National Preparatory School and then the National School of Law, graduating in 1929.

The personal experiences of a president influences his sources of initial political recruitment (see table 5-1). In the case of Cárdenas, the Revolution was central. After all, men under battle conditions develop trust in one another and respect for survival skills. A third of Cárdenas's collaborators had come in contact with him through shared service in the Revolution. Once Cárdenas began a political career, he met other men in the bureaucracy who accompanied him up the political ladder. Although relatively unschooled, when he was governor of his home state, he held weekly seminars for students and professors from the local university, forming close ties with individuals whom he brought into political life later on. Cárdenas served as president of the National Revolutionary Party (PNR), an antecedent of the PRI, but did not recruit from this source. The contrast between him and Alemán could not be more remarkable. Over four-fifths of Alemán's chosen political associates had been classmates or professors at the two schools he had attended in Mexico City.

Alemán established the overwhelming value of preparatory and university education as the institutional locus of Mexican political recruitment, a pattern unchanged to this day. Its importance increased as greater numbers of future politicians began to attend the National Preparatory

Table 5-1 Political Recruitment Sources for Presidents Cárdenas and Alemán

President	Sources of Initial Recruitment					
	Revolution (%)	State (%)	Bureaucracy (%)	Party (%)	School (%)	Relatives (%)
Cárdenas	34	18	26	0	18	3
Alemán	0	10	3	3	85	0

Source: Roderic Ai Camp, *Mexico's Leaders, Their Education and Recruitment* (Tucson: University of Arizona Press, 1980), 22.

Table 5-2 University Graduates by Presidential Administration, 1920–1991

President	Universidad Nacional Autónomo de Mexico (%)	Military (%)	Private (%)	Other (%)
	Institution			
Obregón, 1920–1924	50	9	0	41
Calles, 1924–1928	37	0	5	58
Portes Gil, 1929–1930	33	0	0	67
Ortiz Rubio, 1930–1932	43	21	0	26
Rodríguez, 1932–1934	50	0	0	50
Cárdenas, 1934–1930	27	7	3	74
Avila Camacho, 1940–1946	36	7	4	53
Alemán, 1946–1952	50	5	4	41
Ruiz Cortines, 1952–1958	36	8	1	55
López Mateos, 1958–1964	47	7	1	45
Díaz Ordaz, 1964–1970	51	7	1	41
Echeverría, 1970–1976	54	7	2	37
López Portillo, 1976–1982	52	7	2	39
De la Madrid, 1982–1988	56	5	6	33
Salinas, 1988–1991	51	9	13	27

Source: Roderic Ai Camp, *Mexican Political Biography Project,* 1991.

School and, more significant, the National University. Having attended the former reached an all-time high during the 1958–1964 administration in which 58 percent of the appointees were graduates. Graduates of the National University reached their highest level under President de la Madrid (1982–1988), when they accounted for 56 percent of his college-educated officeholders. Midway through the Salinas administration, National University graduates accounted for half of all national politicians.

The university and preparatory school became important sources of political recruitment for two reasons (see table 5-2). Many future politicians teach at these two institutions, generally a single course. They use teaching in part as a means to identify students with intellectual and political skills, helping them to get started in a public career. Typically, they will place a student in a government internship or part-time job, followed by a full-time position after graduation. De la Madrid is an excellent illustration of this, having started his career in the Bank of Mexico (Mexico's "federal reserve bank") on the recommendation of an economics professor.[6]

Some Mexicans with political interests start out in federal agencies in technical posts or as advisers. As their interest in politics grows, they develop contacts with other ambitious figures, including an agency superior. That individual, like the politician-teacher, will initiate their rise within the national bureaucracy. Today the federal bureaucracy ranks second only to the university as a source of political recruitment, and grows in usefulness, once the individual is launched in politics.

In the United States, Democratic and Republican party organizations are often the source of nationally prominent politicians. The parties carry much more weight in the recruitment of politicians because of their role in the candidate-selection process, and in the competition for offices that influence policy-making a great deal. In other words the American politician has to impress not only the electorate but also local, state, or national party leadership in order to obtain the party nomination.

In Mexico few *national* politicians are recruited through party channels, especially at the local and state levels, although many prominent figures at the state and local levels are recruited in such fashion. Because executive-branch leadership controls the party, and decision making is centralized in the executive branch rather than in the legislative bodies, a career in the national bureaucracy is the foremost means of ascent. Among national officeholders in the Salinas administration, only a minuscule 2 percent served in a local government post. In recognition of this fact of life, budding politicians descend upon Mexico City, there to carve out careers in the federal government.

The salience of the federal bureaucracy in the recruitment process contributes to another informal characteristic of upward political mobility in Mexico: the significance of Mexico City. Politicians who come from Mexico City, in spite of its tremendous size, are overrepresented in the national political leadership. This was true prior to the Revolution of 1910, but those violent events introduced a leadership whose birthplaces deemphasized the importance of the capital. That remained true until the 1940s, when the presence of Mexico City in the backgrounds of national politicians increased substantially. By the presidency of Luis Echeverría (1970–1976), when fewer than one in ten citizens was born in Mexico City, one in four national political figures named it as a place of birth. In the past twenty years, that figure increased dramatically: Mexico City is the birthplace of nearly half of president Salinas's appointees who are holding national office for the first time, nearly four times that of the general population of the same age. These figures suggest that for a variety of reasons, growing up in Mexico City is a tremendous advantage to the politically ambitious.

In all political systems, whom one knows has much to do with political recruitment and with achieving an important political office. U.S. politics is replete with examples of prominent figures who have sought out old friends to fill responsible political offices. In fact, knowing someone is often a means for obtaining employment in the private sector as well. In

Mexico, whom one knows in public life is even more telling, given the fact that incumbent officeholders often have the say on who obtains influential posts. Mexicans with political ambitions can enhance personal contacts at school, in the university, or during their professional and public careers through family ties.

Americans have produced a few notable political families, say, the Adamses and Kennedys, and George Bush is the son of a U.S. senator, but such families are numerous in Mexico. One reason, as studies of British and U.S. politicians have shown, is that children of political activists are more likely to see politics as a potential career than are children reared in a nonpolitical environment.[7] It is natural that a youngster growing up in a political family would come in contact with many political figures. More than one in eight Mexican national politicians from 1970 through 1988 were the children of nationally prominent political figures. President Salinas himself is such a one. His father, who served in the cabinet in the 1960s, helped his son's early career. If extended family ties are considered, between one-fifth to one-third of all politicians were related to national political figures during the same period.

Politically active families are not the only factor that makes family background important. Social and economic status is another. Studies of politicians worldwide, in socialist and nonsocialist societies, suggest the importance of middle- and upper-middle-class backgrounds.[8] In Third World countries without competitive political structures, family origins become even more significant to career success.

Higher socioeconomic backgrounds are helpful in political life because well-off parents provide opportunities for their children. Education, as we have seen, is such a significant means of political recruitment, hence *access* to it enables making the right contacts and obtaining the necessary informal credentials. Some Mexicans from working-class backgrounds manage to attend preparatory school and even a university, but few actually complete degree programs. Financial ease contributes to acquisition of a degree if desire and intellectual capacity are present. This explains why Mexico's youngest generation of national politicians, those born since 1940 (President Salinas's generation), are almost exclusively from middle- and upper-middle-class backgrounds.

Another informal credential universal to political leadership in all countries is gender. Politics has been, and remains, dominated by men. Nevertheless, women have made substantial inroads in national political office. On the whole, women have been far more successful politically in Mexico than in many other countries, including the United States. For example, several women served on the supreme court in Mexico, long before Sandra Day O'Connor was appointed by President Reagan. Numerous women have held Senate positions. Top cabinet posts have occasionally been filled by women, but men have a virtual lock on that domain, especially in the major agencies.

Slightly different recruitment patterns have traditionally been followed by women interested in politics. This fact works against their obtaining the higher positions because they do not come in contact with current and future political figures who could assist them up the bureaucratic ladder. Typically, women politicians have been found more frequently in party posts and in the legislative branch, and they have not obtained the same type or level of education. Younger women who are politically ambitious are now taking on many of the characteristics of their male peers. Among younger political figures (born after 1950) at the national level, women now account for one in four individuals. The representation of women in national political offices (cabinet, subcabinet, and top judicial and legislative posts) increased substantially after 1976, during the administration of José López Portillo (see table 5-3). The lower figure for Salinas does not necessarily represent a decline in the representation of women; the data are for only the first half of his administration.

Successful political recruitment in Mexico requires certain informal credentials. Higher education is an essential; higher education in Mexico City, and at the National University, is extremely advantageous. Second, it is very helpful to a political career to have been born in the capital, which strengthens the potential for attending the universities from which most establishment politicians have graduated. Third, young professionals who join the federal bureaucracy rather than the party bureaucracy or state and

Table 5-3 Women's Recruitment to National
Political Office by Administration, Mexico,
1935–1991

President	Percentage of Women in Administration
Cárdenas, 1934–1940	0
Avila Camacho, 1940–1946	1
Alemán, 1946–1952	2
Ruiz Cortines, 1952–1958	4
López Mateos, 1958–1964	4
Díaz Ordaz, 1964–1970	6
Echeverría, 1970–1976	8
López Portillo, 1976–1982	19
De la Madrid, 1982–1988	17
Salinas, 1988–1991	11

Source: Camp, *Mexican Political Biography Project,* 1991.

local agencies are much more likely to reach the top. Fourth, a middle-class background is almost essential to achievement of the education necessary for a public career. Finally, family ties to successful politicians have enhanced the career opportunities of many of Mexico's leading public figures; all the presidents since 1970 have been related to prominent political figures.

THE CAMARILLA: GROUP POLITICS IN MEXICO

Perhaps the most distinctive characteristic of Mexican politics, knowledge of which is essential to understanding the recruitment process, is the political clique, the *camarilla*. It determines, more than any other variable discussed, who goes to the top of the political ladder, what paths are taken to do so, and the specific posts they are assigned. Many of the features of Mexican political culture predispose the political system to rely on camarillas. A camarilla is essentially a group of individuals who have political interests in common and rely on one another to improve their chances within the political leadership (see figure 5-1).

A camarilla is often formed early in several individuals' careers, even while they are still in college. Its members place considerable trust in one

---

Camarilla: a group of individuals who share political interests, and rely on one another to improve their chances within the political leadership.

---

another. Using a group of friends to accomplish professional objectives is a feature found in other sectors of Mexican society, including intellectual life and the business community. A camarilla has a leader who acts as a political mentor to other members of the group. He typically is more successful than his peers, and uses his own career as a means of furthering the careers of other group members. As the leader of a camarilla ascends in the bureaucracy, he places members of his group, when possible, in other influential positions either within his agency or outside it. The higher he rises, the more positions he can fill.[9]

Because it has retained a monopoly for more than sixty years, Mexico's political leadership can be viewed as overlapping and hierarchical camarillas, all linked. They are fluid groups, and if a mentor's career does not advance, it is acceptable to shift loyalties to another camarilla. It is also permissible to have ties to more than one camarilla, although at a given time, one is identified specifically with only a single group.

Figure 5-1 Fifteen Characteristics of Mexican Camarillas

1. The structural basis of the camarilla system is a mentor-disciple relationship that has many similarities to the patron-client culture throughout Latin America.

2. The camarilla system is extremely fluid, and camarillas are not exclusive but overlapping.

3. All successful politicians are the products of multiple camarillas, that is, rarely does a politician remain within a single camarilla from the beginning to the end of his or her career.

4. Mexicans who successfully pursue politics as a profession initiate their own camarillas simultaneously with membership in mentors' camarillas.

5. Every major national figure is the "political child," "grandchild," or "great grandchild" of an earlier, nationally known figure.

6. The larger the camarilla, the more influential its leader and, likewise, his disciples.

7. Most significant camarillas today can be traced back to two major political figures: Lázaro Cárdenas and Miguel Alemán.

8. Some camarillas are characterized by an ideological flavor, but other personal qualities generally determine disciple ties to a mentor.

9. Disciples often surpass the political careers of their mentors, thus reversing the benefits of the camarillas relationship and the logical order of camarilla influence.

10. Camarillas formed largely within an institutional environment, have become increasingly significant as decision making, especially in the economic realm, has become more complex. The single-most-important public institution representative of this trend, especially relative to its size, is the Bank of Mexico.

11. Kinship and educational companionship are the major sources of camarilla loyalties today, but professional merit, contrary to popular assumptions, has become increasingly important.

12. All politicians automatically carry with them membership in an educational camarilla, represented by their preparatory, professional, and graduate school generation.

13. Politicians with kinship camarillas have advantages over peers without such ties.

14. Politicians who are most adept at building camarillas on the basis of professional merit have the largest and most successful groups over time but are not necessarily the most likely to achieve the presidency.

15. Because of the overlapping quality of camarillas, some politicians have shared loyalties. Normally, most politicians, at a given time, can be identified with a specific camarilla. It is acceptable to shift loyalties when the upward ascendancy of the political mentor is frozen.

Source: Roderic Ai Camp, "Camarillas in Mexican Politics: The Case of the Salinas Cabinet," *Mexican Studies* 6, no. 1 (Winter 1990): 106–7.

Mexican politics, from the postrevolutionary generation forward, has been built upon the interrelationships of the camarillas. Indeed, all contemporary camarillas have their origins in two major forebears: the camarillas of Cárdenas and Alemán. Cárdenas's personal camarilla spawned four successive generations of camarillas, accounting for at least 144 national officeholders. Cárdenas's most important disciple today is his own son, Cuauhtémoc, who took the unusual step of leaving the establishment leadership in 1987 to form his own party. He ran for the presidency in 1988, winning the largest percentage of presidential votes ever recorded for an opposition party in modern Mexican history.

The interrelatedness of the camarillas is reflected by the fact Alemán was a disciple of Cárdenas's. Nevertheless, it was Alemán who altered the nature of political recruitment as exercised by his mentor, emphasizing, as suggested above, the importance of educational and bureaucratic contacts over Cárdenas's relationships derived from the Revolution.

The camarilla takes on added importance in politics because the mentor establishes the criteria by which he chooses his disciples. The most successful camarilla reaches the presidency, thus influencing the entire system. The implications for political recruitment are crucial. It has been shown that politicians, like most people, tend to recruit individuals with similar credentials or experience, who in many ways mirror themselves.[10] Over time, incumbents can structure the recruitment process to favor certain credentials. Generally, however, presidents, because they exercise the most comprehensive influence over political appointments, have the greatest impact on the recruitment process.

It has been pointed out that Alemán introduced some new credentials because of the individuals he chose for high political office, giving his peers the opportunity to reinforce those same credentials. From the 1940s through the 1970s, as the camarillas introduced by Alemán and his generation rose to the top, certain credentials became increasingly important: a college education, preferably from the National University; an urban birthplace, preferably Mexico City; a career in national politics, preferably the federal bureaucracy; pursuit of a law degree and legal career; and entrance into public service at a young age, often while still in college.

In the 1980s, and in some instances earlier, a change began to occur in the informal credentials required of the most successful politicians. A policy setting forth these changing requirements was not established; rather, they emerged naturally as the politicians themselves changed their credentials. These recent trends have become sharper and more easily recognized under Salinas.

The three most important sources of contemporary political camarillas in Mexico are family, education, and career. Family has remained consistently at the fore, and has changed only in the sense that today's politicians are increasingly the children of national political figures. In the past, family relationships have been frequent but not as direct.

Educational and career characteristics of politicians have altered markedly in the past two decades. Educationally, the most persistent change has been the constant increase in *level* of education. Not only are all national political leaders, with a few exceptions in the legislative branch, college educated but graduate education has reached new highs. Both de la Madrid and Salinas obtained graduate degrees. De la Madrid has an M.A. degree in public administration from Harvard; his disciple and successor, Salinas, has two M.A. degrees as well as a Ph.D. degree from Harvard. These two presidents reflect the importance given to advanced education in Mexican politics. Of the *new* national officeholders under de la Madrid, nearly half, like the president, had graduate degrees. Only six years later, beginning with the Salinas administration in 1988, 70 percent have received graduate training; many of them are Ph.D. degree holders.

De la Madrid and Salinas introduced another informal credential in the recruitment process. Their camarilla selections emphasized politicians who had been educated outside Mexico, particularly at the graduate level. It can be said that Salinas was following in the footsteps of his father, who also graduated from Harvard with an advanced degree. The point is that numerous political figures began to study abroad, generally at the most prestigious universities in the United States. In Salinas's administration Harvard and Yale graduates are most typical.

A third alteration in the educational background of contemporary politicians, and perhaps the most significant, is the discipline studied. Law, as in the United States, has always been the field of study of most future politicians; engineering and medicine, secondary. This means that law school is the most likely place to meet future politicians and political mentors. The most remarkable change is from the primacy of law to economics. Salinas is the first president with that specialty, and his political generation is the first to count as many economists as lawyers among its members.

The new discipline emphasis has led to another significant change in recruitment characteristics: the elevation of private over public education. This characteristic is less pervasive than the others but even more remarkable. Between the administration of de la Madrid and Salinas, a sixfold increase in the percentage of private-school graduates has taken place. Instead of the National University and public universities maintaining their level of dominance, private institutions have begun to make serious inroads. This is significant for political recruitment because it will change not only informal credentials, but the locations where recruitment takes place. In fact, it contributes to the diversity of the recruitment process, a process which traditionally has relied on fewer educational sources.

THE RISE OF THE TECHNOCRAT

As the recruitment process changed, and the credentials of future politicians were modified, some scholars labeled the younger generation of politicians in Mexico as technocrats, *técnicos*. The rise of technocratic leadership took place throughout Latin America. A number of attributes have been associated with this class of leaders in Brazil and Chile, and many have been mistakenly applied to Mexico's leaders. This has generated some confusion about technocrats.

Mexico's technocratic leadership is characterized by new developments in their informal credentials; in particular, they are seen as well educated in technologically sophisticated fields, as spending most of their careers in the national bureaucracy, as having come from large urban centers, notably Mexico City, and from middle- and upper-middle-class backgrounds, having studied abroad (see figure 5-2). By implication, in

Figure 5-2 Characteristics of Mexico's Politician-Technocrat, 1990s

Characteristic	Percentage Having
Urban birthplace	94
Male	87
Middle-class parents	85
College educated	83
Graduate of the National University	58
Taught	57
Prior national political post	56
Born between 1920 and 1939	52
Graduate education	46
Lawyer	37
Taught at the National University	37
Graduate of the National Preparatory School	29
Ph.D. degree	20
Graduate work in the United States	19
Economist	16

Source: Mexican Political Biography Project, 1991.

contrast to the more traditional Mexican politician, they have few direct ties to the masses, and in terms of career experience, lack elective officeholding and grass-roots party experience.

For some years the typical Mexican politician has been a hybrid, exhibiting characteristics found among *técnicos* and traditional politicians. The assertion by some scholars that technocrats lack political skills is incorrect and misleading. Technocrats as a group do not have an identifiable ideology. The political-technocrat, a more apt label, is primarily distinguished from the politician of the 1960s or 1970s by lack of party experience, by the fact that he or she has never held elective office, and by specialized education abroad. These characteristics, for example, are found in President Salinas's own career (see figure 5-3). The implication of these three characteristics is that the politician-technocrat, although highly

Figure 5-3 Career Progression of Carlos Salinas Gortari, Technocrat President

1982–1987	Secretary of Programming and Budgeting
1981–1982	Director General of the Institute for Economic, Political and Social Studies (IEPES), PRI, presidential campaign
1979–1981	Director General of Social and Economic Policy, Budgeting
1979–1981	Technical Secretary of Economic Cabinet
1979–1979	Subdirector, Economic Studies, IEPES, PRI
1978–1979	Director General of Treasury Planning, Treasury
1978–1978	Subdirector General of Treasury Planning, Treasury
1978–1978	Professor, fiscal policy, Center for Monetary Studies of Latin America (CEMLA)
1976–1978	Ph.D., political economy, Harvard University
1976–1977	Director of Economic Studies, Treasury Planning
1976–1977	Technical Secretary Internal Groups, Treasury
1976–1976	Subdirector, Economic Studies, Treasury Planning
1976	M.A., political economy, Harvard University
1976	Professor, public finance, Technological Institute
1974–1976	Head, Financial Studies and International Affairs, Treasury
1974	Research assistant in public finance, Harvard University
1972–1973	M.A., public administration, Harvard University
1971–1974	Adviser, Subdirector of Public Finance, Treasury
1971–1971	Teacher, Institute of Political Education, PRI
1970–1972	Adjunct Professor, economics, National University, Mexico City
1966–1969	Economic studies, National University, Mexico City
1966–1968	Aide to President of PRI, Federal District
1963–1966	Preparatory School #1 San Ildefonso, Mexico City
1960–1963	Secondary School #3 Heroes de Chapultepec, Mexico City
1953–1959	Abraham Lincoln School (private), Mexico City

skilled, does not possess the same political bargaining skills as the peer who has had a different career track, and that such an individual *may* be more receptive to political and economic strategies used in other cultures as a consequence of foreign education. For example, some critics suggest that Salinas's economic cabinet, whose members have these technocratic characteristics, welcomed the economic liberalization philosophy of western Europe and the United States because of their economic background and education abroad.

CONCLUSION

The formal structure of Mexico's political system sheds little light on how interested Mexicans pursue successful political careers. The political recruitment process is strongly affected by the centralization of political authority, and the reality of incumbent selection. Because other variables and groups exercise little influence over the appointment and "election" of government officials, informal credentials have replaced more formal requirements as helpful or essential to political recruitment.

The informal credentials have been associated with generations of political leaders since the 1920s. As incumbent leaders have changed their own credentials, they have passed on the changes to succeeding generations of politicians. Contrary to expectations, and different from the United States, educational institutions take on very important roles in political recruitment. Many Mexican politicians teach at the university level, and use their classes as a means of recruiting potential politicians. The most important governmental source of political recruitment is the national bureaucracy, not the legislative branch. Mexico's centralization of authority also contributes to the importance of national institutions, and to the neglect of local and state institutions in the backgrounds and career experiences of politicians.

The political clique, the camarilla, plays the most important role in the recruitment process. It is an informal structure built on several characteristics of the general culture, a structure emphasizing the importance of placing career loyalties in the hands of close friends, and using a group of friends to enhance one another's career success. All prominent Mexican politicians, excluding the opposition, are members of camarillas and are tied to one another through fluid linking among the many cliques. Contemporary politicians can trace their camarillas back to those originated by Cárdenas and Alemán.

The most salient characteristics of a successful contemporary politician, reflected in the background of President Carlos Salinas himself, include high level of education, graduate training abroad, a degree in economics or a more technically specialized discipline, a middle-class social origin, a career in the national bureaucracy, especially economics-oriented agencies like treasury, residence in an urban center, especially Mexico City, and, increasingly, graduation from a private, not a public, institution in the capital.

The changing features of the recruitment process and the changing characteristics of politicians themselves enabled the rise of a new politician in Mexico, commonly labeled a technocrat. Salinas is a prototype. Politician-technocrats have increasingly dominated the government, especially the executive branch, and control Salinas's economic cabinet. Several phenomena have been attributed, often incorrectly, to their personal qualities. If Salinas's policies are deemed worthwhile, politician-technocrats are likely to dominate the government through the end of the century.

NOTES

1. Lester G. Seligman, *Recruiting Political Elites* (New York: General Learning Press, 1971).

2. Kenneth Prewitt, *The Recruitment of Political Leaders: A Study of Citizen-Politicians* (Indianapolis: Bobbs-Merrill, 1970), 13.

3. See Ralph Turner, "Sponsored and Contest Mobility and the School System," *American Sociological Review* 25 (December 1960): 855–56.

4. See Robert D. Putnam, *The Comparative Study of Political Elites* (Englewood Cliffs, N.J.: Prentice-Hall, 1976), 45ff.

5. See Roderic A. Camp, "Mexican Presidential Candidates: Changes and Portents for the Future," *Polity* 16 (Summer 1984): 588–605.

6. Personal interview with Miguel de la Madrid, Mexico City, 1991.

7. See, for example, Richard Rose's statement that "the number of politicians from political families is disproportionately high in every Cabinet." *Politics in England: Change and Persistence,* 5th ed. (Boston: Little, Brown, 1989), 177. For the United States, see Alfred Clubok et al., "Family Relationships, Congressional Recruitment, and Political Modernization," *Journal of Politics* 31 (November 1969): 1036.

8. Thomas R. Dye, *Who's Running America? Institutional Leadership in the United States* (Englewood Cliffs, N.J.: Prentice-Hall, 1976), 152; George K. Schueller, "The Politburo," in *World Revolutionary Elites,* ed. Harold D. Lasswell and Daniel Lerner (Cambridge: MIT Press, 1966), 141.

9. For an excellent description of this process, see Merilee S. Grindle, "Patrons and Clients in the Bureaucracy: Career Networks in Mexico," *Latin American Research Review* 12, no. 1 (1977): 37–66.

10. Kenneth Prewitt and Alan Stone, *The Ruling Elites: Elite Theory, Power, and American Democracy* (New York: Harper & Row, 1973), 142.

6

Groups and the State:
What Is the Relationship?

> Formally, Mexico has a corporatist system of interest represen-
> tation, in which each citizen and societal segment must relate to
> the state through one structure "licensed" by the state to
> organize and represent that sector of society (peasants, urban
> unionized workers, businessmen, teachers, etc.). The official
> party itself is divided into three sectors: the peasant sector, the
> labor sector, and the "popular" sector (representing most gov-
> ernment employees, small merchants, private landowners, and
> low-income urban neighborhood groups). Each sector is domi-
> nated by one mass organization. . . .
>
> WAYNE A. CORNELIUS AND ANN CRAIG, *Politics in Mexico*

All political systems, regardless of whether the struggle for political power
is highly competitive or strongly monopolized by a small leadership group
or single party, must cope with political interests and groups. In the United
States various interest groups, as they are labeled, express their demands to
the executive and legislative branches and contribute significant sums of
money to parties and candidates. In Mexico, given the political system's
structural differences from that of the United States, both the type of
groups and their means for influencing public policy are unlike what obtains north of the border.

THE CORPORATIST STRUCTURE

The importance of corporatism to the Mexican political culture and model
was suggested earlier.[1] Corporatism describes the more formal relationship

between selected groups or institutions and the government or state. Since the Revolution, that is, for most of the twentieth century, Mexico has been using an interesting structure to channel the most influential groups' demands that has enabled monitoring the demands and mediating among them. The government has sought to act as the ultimate arbiter and to see to it that no one group becomes predominant.

The corporatist structure was largely devised and put in place under President Lázaro Cárdenas (1934–1940). Although Cárdenas wanted to strengthen the state's hand in order to protect the interests of the ordinary worker and peasant, he ironically created a structure that for the most part has benefited the interests of the middle classes and the wealthy, not unlike that of many other political systems.[2] The reason for the outcome is that the commitment of Cárdenas to the social welfare of the less well off has not been shared by most of his successors, who have responded to other concerns and groups.

What is important, however, is that although the ideological orientation has changed and various economic strategies have been experimented with since the 1930s, the arrangement remained largely intact until the 1990s.[3] Only under President Salinas has there been some interest in restructuring the corporatist relationship[4]—this in response to Salinas's promises of political modernization and democracy. Observers argue that corporatism contradicts democracy, and that the greatest stumbling block to a functioning Mexican democracy is the continuation of corporatist structures of Cárdenas and his successors.[5]

The corporatist features of the political system allow two types of channels for making political demands and consequently two types of institutional representatives have arisen. The institutional relationship with the government under this type of system is traditionally a formal one: the state establishes an organization, requiring individuals meeting the criteria of a special interest to belong. For example, the state created several business organizations to which businesses employing a certain number of employees must belong.[6] The state, however, has not always managed to control all institutions representing various groups. Those it controls are considered to be quasi-governmental interest organizations. As some interest groups' influence grew and they became more autonomous, they created their own organizations, considered to be independent or autonomous. For example, the business community established the Mexican Association of Employers (Coparmex), an influential private-sector voice.

The other kind of channel is informal channels, which are characteristic of all government models. Certain groups in Mexico do not use formal institutions, independent or governmental, to exercise their considerable influence but instead use informal channels. The informal channels may be incorporated within the governmental structure or remain independent of it. Although it cannot be asserted with complete certainty that the informal channels are more significant than the formal channels, given the lack of relevant studies, most observers of Mexican politics believe that to be the case.

INSTITUTIONAL VOICES

The range of interest groups in the United States is formidable because of the political system's openness and the ability of multitudes of like-minded citizens and institutions to organize. Such collectivities in Mexico are few and weak, and do not figure in decision making.[7] Remember that decision making is centered in the executive branch, thus blocking the ability of diverse interests to pressure the legislative branch. Further, the prohibition against running consecutively for legislative seats limits the potential threat that interests can level against individual members of Congress.

The most important groups incorporated formally and informally in the corporatist structure are the military, the Catholic Church, business, organized labor, and intellectuals. Each has a somewhat different relationship to the government institutionally.[8] Most have stood out historically in other Latin American countries, suggesting the significance of similar past experiences and the influence of the colonial heritage on contemporary politics in the region.

The Military

No group has played a more significant role in Latin American political life than the military. However, its pattern of influence in Mexico has been quite different since the 1930s from that found elsewhere in Latin America. Most important, the military has found it necessary to intervene politically in every Latin American country except Mexico since that decade, and in most countries the military seized power in the 1970s and 1980s.[9]

The military's relationship to the Mexican state or government is different from that of most other groups. The reason is that the military does not function as a separate political actor; rather, it is part of the government apparatus and operates under civilian leadership. This is not to suggest that the military does not have institutional interests, that is, that it does not have its own interests that differ from those of the political leadership and other interest groups.

Since the 1930s Mexico's civil-military relationship has been increasingly characterized by subordination of the military, and of its interests to those of society as defined by the civilian leadership.[10] Aside from Costa Rica, which has operated without an army since the late 1940s, and Venezuela, which began subordinating the military to civilian rule in the late 1950s, Mexico is an exception. How did its unusual relationship come about?

When Cárdenas became president, and he was part of the generation that had participated in Mexico's civil war, he incorporated the military into the recently established government party, the National Revolutionary party (PNR). He wanted to balance the military against the agrarian and labor sectors within the party, and thus lessen its overall political influence.[11]

Cárdenas's successor in the presidential chair, General Manuel Avila Camacho, altered this major characteristic of the early corporatist structure by removing the military as a separate party sector.[12] Basically, he did not want to recognize the military as having a public political voice, and did not want to give it equal standing with other notable interest groups. From the 1940s to the present, then, the military's relationship to the government has been determined by its formal structural ties to the executive branch, and through informal channels.

The political leadership gradually reduced the military's political influence through a variety of techniques. In the first place, as has been shown by James Wilkie, each successive government reduced the military's allocation as a percentage of the federal budget;[13] the amount increased slightly each year, but the *percentage* went down. The size of the military in relation to population, and the sum budgeted to the military per capita, is among the lowest worldwide, far below the figures for the United States.[14]

As the political leadership gradually reduced the size and potential influence of the military, it strengthened the legitimacy of political institutions, including the official party. The leadership had the advantage of operating in a semiauthoritarian fashion within the electoral arena. Military intervention is generally facilitated by competing political groups in a society that are seeking allies in the military. In Mexico, however, the military had to be either for the establishment—that is, the civilian leadership—or against it, and had no outside civilian allies after 1952. During the 1940s, 1950s, and 1960s, military officers who pursued political careers helped bridge the gap between the civilian and military leaderships. In other words, these political military officers provided a significant, *informal* channel of communication, allowing the civilian leadership to solidify its control and to establish its legitimacy.

Civilian leadership also cemented its control over the military through the professional socialization process. Civilian politicians established several military schools, most notably the Heroic Military College, the Superior War College, and in 1981 the National Defense College, to train officers. One of the most important themes in the curriculum of the schools is respect for authority, for one's superior officer, and for the commander in chief, the president. Although all military schools tend to drill in the concept of subordination to authority, Mexican military academies are famous for the level of discipline they instill. An American officer, a graduate of a U.S. military academy, writes, "First, the dominant value would be the individual's willingness to subordinate himself totally to those in authority over him. Coupled with this would be the expectation that submission will be rewarded and independence will be severely punished. Finally, the officer's primary motivation would be to secure the rewards that the system has to offer."[15]

For the above reasons and many others, the military is clearly subordinate to the civilian political leadership in Mexico. This does not mean that it has little or no influence on the government. The military has served the government in many capacities other than those traditionally subscribed to by the military in the United States.[16] The Mexican military's primary responsibility has not been national defense; rather, it has operated in many realms as an internal police force devoted to national security.[17] Not only does it provide the government with political intelligence, it has been used to maintain electoral peace, to settle contentious strikes, and in the 1980s and 1990s to carry out antinarcotics raids.

The Church

Although Mexico legally established, and in practice enjoys freedom of religion, Mexicans, as we noted in chapter 4, are overwhelmingly Catholic, products of a Catholic, Christian culture.[18] The Catholic Church exercised extraordinary political influence in Mexico and elsewhere in the region during the colonial period and continued to do so in the nineteenth century and part of the twentieth.[19] The pattern was broken at about the same time that the government began to reduce military influence significantly. The restrictive provisions in the 1917 Constitution were implemented.

The Church, unlike the military, operated as an institution fully independent and autonomous of the government, yet severely hampered in theory and practice by the Constitution. The state reached an informal understanding with the Church after 1930 that in effect allowed the Church to carry out the spiritual and pastoral functions within the purview of all churches in return for its remaining quiet about political and social issues. The understanding remains in effect to this day, and in practice was fairly well adhered to by both parties until the early 1980s.

The Church's role as an interest group is limited because of the anti-Church rhetoric that is incorporated into the public education of each child in Mexico. The Church and clergy were at a disadvantage compared to some other groups in the corporatist arrangement because of the legal limbo they occupied. For example, the Church as an institution had no legal standing until 1992, the only institution of the five sectors under discussion so characterized. Clergy of all faiths did not have the legal right to vote before 1992, although many actually did so.[20]

As chapter 4 relates, Mexicans nonetheless remain very religious; many are practicing Catholics, and most have a high regard for clergy and the Church as an institution. Because respect for the Church is high, and because opposition political organizations have not provided viable channels for people to express their political demands, some Mexicans have turned to the Church for guidance—and more important, as an institutional vehicle to convey their political frustrations.

The Church, as is true of other groups, such as businessmen, does not speak with a single voice. Despite its image as a centralized, hierarchical institution, it is decentralized at the level of individual dioceses, of which there are sixty-nine in Mexico. Diocese and archdiocese are territorial subdivisions that serve as organizational units, each is governed by a bishop or archbishop. Collectively, these men are the extremely autonomous hierarchy of the Church. In recent years numerous bishops have spoken out publicly on issues affecting their dioceses.

It is apparent from recent events in Mexico that geography and the social and economic composition of a diocese often affect the attitude and orientation of its priests and bishops. One of the issues that has increasingly disturbed some bishops is electoral fraud. Many bishops believe it is their responsibility to take stands on matters of social and political consequence, a belief that has eventuated in both implicit and explicit criticism of government actions.[21] One bishop commented:

> One point that is important to make which is independent of the present condition in Mexico is that the Church, both the bishops and priests, consid-

er it necessary to socialize the people about their civic obligations regardless of what the political situation might be. The people are very ignorant of their civil responsibilities. From a moral point of view, we need to create a sense of consciousness. All of this can be badly interpreted by the government, which may see us wanting to reestablish political privileges we have had historically. For us, however, it is obvious that we have no desire to make policy decisions that are handled presently by the government. We only believe we have a responsibility to defend the people. Who else is there?[22]

Some bishops have strengthened their positions by joining together to explain their positions. The most memorable instance in the past decade was that of the northern bishops, led by Archbishop Adalberto Almeida of Chihuahua; they condemned election fraud in Chihuahua in 1986 and called on the administration of Miguel de la Madrid to annul the results and hold new elections.[23] The bishops threatened to withhold the celebration of masses until the government responded to their demands. Although the pope intervened to prevent their carrying out the threat, their public posture in direct violation of the Constitution illustrated their potential influence.

The leadership of the Church in Mexico in terms of policy influence is the episcopate—the body of bishops, archbishops, and cardinals—which in conference recommends policies on issues ranging from the purely theological to foreign debt, the maldistribution of income, and drugs. The episcopal meetings result in the publication of pastoral letters and enunciations of recommended positions.

President Salinas has moved, as part of his modernization plans, to make the Catholic Church a more open actor in the political system. Although many politicians are resistant to any changes in the constitutional restrictions on the Church, Salinas believes the relationship is outdated and must be refashioned. Indicative of his new posture, he invited leading clergy to attend his inauguration in December 1988, and then appointed a former political figure as his personal representative to the Vatican. He also made Pope John Paul a welcome guest in Mexico in the summer of 1991, furthering even closer relations between the government and the Church. By 1992, he had revised several major constitutional provisions, one of which now permits recognition of all churches as legal entities.

The Church generally does not openly lobby for its political positions; rather, it requests, and receives, audiences with the state officials. Typically, party presidential candidates meet with bishops during their campaigns.[24] Church personnel also meet with various members of the executive branch on matters of mutual concern. On the state level, bishops frequently meet with state governors and collaborate with the government on social welfare projects. Although relations are good as a rule, and have improved considerably since 1989, at the local level in certain instances, such as in the southern state of Chiapas in 1991, they may be quite conflictual. In this particular instance, the governor attempted to have an abortion law passed in 1990, and a year later, arrested and imprisoned a priest for allegedly having led a peasant group to occupy privately owned lands.[25]

Business

The private business sector combines some of the corporatist features of a governmental institution, the military, with those of an autonomous institution, the Church. As suggested previously, it has an array of organizations that present its demands to the government. The most important quasi-governmental organizations, established by the government itself, are a group of federations that include the National Chamber of Industries (Canacintra, or CNIT), the National Chamber of Commerce (Concanaco), and the National Federation of Chamber of Industries (Concamin). The most important autonomous organizations, in addition to Coparmex, are the Mexican Insurance Association (AMIS), the Mexican Council of Businessmen (CMHN), and before the nationalization of the banks in 1982, the Mexican Bankers Association.

When Cárdenas established some of these quasi-governmental business groups, the private sector was rather weak. As it has grown, it has not only developed other organizations to represent its own interests but often taken positions on economic policies different from those advocated by the government.[26] The private sector, however, has labored under conditions similar to the constraints on the Church, although not nearly as extreme. The government has allowed labor, professional organizations, and peasants to be formally represented in the party, but it has purposely excluded the private sector. It has done so because private-sector interests have not coincided with the rhetoric of the postrevolutionary leadership, even if in reality interests have been shared.

Business groups do not easily fit into the corporatist model used by some scholars to describe Mexico's political system. This is so because the quasi-governmental organizations are not the most important means for expressing private-sector demands. Again, the significance of *informal* means to express those demands becomes apparent. One prominent businessman described the actuality:

> Sometimes it is the business groups which approach the government concerning policy questions, and in other situations it is the government which takes the initiative with the private sector through the individual chambers. It

is really what you might call a corporatist situation in which the government and the private sector are tied together as far as interest representation. The difference between our system and that in your country is that here we try to influence directly the minister of the appropriate secretariat rather than going through the legislative branch. Normally, even though we try to directly influence the minister in charge, we first go through the chamber before approaching the individual personally.[27]

Recognizing the advantage of collective representation, at least on certain issues, businessmen created a unitary body to represent the top organizations: the Businessmen's Coordinating Council (CCE). The CCE, however, is not representative of its own members even though it speaks for them. By far the most important business organization is the semisecret CMHN, which is made up of some thirty-eight prominent capitalists. The members meet frequently with not only cabinet members but the president. Yet, as members of the CMHN have revealed, rarely does it make demands on the government or the president; rather, membership in the elite organization is the instrument by which individual access to the president or the appropriate government official is gained.

Unofficial organizations have also exerted some influence. Until the 1970s business and government maintained a relatively stable and symbiotic relationship, although tensions existed.[28] By the end of Luis Echeverría's administration in 1976, the tensions were increasing as the government began to expand its economic role, buying up privately operated enterprises and initiating policies that ran counter to private-sector interests. This culminated in the 1982 decision of President José López Portillo (1976–1982) to nationalize privately owned banks. There followed a significant break between the private sector and the government, and a high level of distrust on the part of the former toward the latter.[29]

President Miguel de la Madrid worked assiduously in the 1980s to repair the damaged relationship and partially succeeded. Nevertheless, smaller independent business groups under Coparmex's vociferous leadership advocated a more energetic political activism for businessmen, including open support for opposition parties. These groups began to campaign for candidates of the National Action Party, and members even ran for state and local offices, especially in northern Mexico.[30] Their position was symbolized in the 1988 presidential race when a successful northern businessman, a former president of both Coparmex and the CCE, opposed Salinas on the PAN ticket.

Businessmen, more than any other group in Mexico with the exception of the political leaders themselves, have the capacity to influence government decisions, but especially in the economic realm, they have not been able to do so consistently or to a meaningful degree. Government economic policy, which has favored their interests more frequently than those of organized labor, has emerged as much from the self-interest or preferences of government leaders as from private-sector pressures.

Organized Labor

Of all the groups with political influence in Mexico, organized labor best meets the criteria of an ideal, corporatist group. One of the contributing causes of the Mexican Revolution was the suppression of the working class under the porfiriato. General Obregón recognized the political importance of labor, and relied on labor's support in his struggles against President Venustiano Carranza. The labor movement started to grow in the 1920s, and in the next decade membership in labor organizations reached 15.4 percent of the economically active work force. It has not grown in percentage terms since 1940, and by 1970 began to decline.[31]

Organized labor in Mexico is quite different from that in the United States. The first distinguishing characteristic is the preponderance of government employees, most of them federal, who account for more than a third of all organized workers. The second characteristic is that organized labor is made up of unions called confederations, similar to the chambers of the business organizations. Nearly half of organized laborers are members of these broad confederations. The third differentiating characteristic of labor is the lesser presence of purely industrial-based unions such as those of miners, electricians, and petroleum workers.

The most important labor organization, the Mexican Federation of Labor (CTM), was established under President Cárdenas. He and his successors maintained a close relationship with union leaders that amounted to government control.[32] The control was cemented by incorporating the CTM as the foundation of one of the three sectorial pillars of the National Revolutionary Party—a role that continues to this day. The CTM has been led since the 1940s by Fidel Velázquez, giving him considerable stature and influence in the labor movement.[33] Organized labor's status within the party, more than any other characteristic, places it in the corporatist fold. Unlike business, the Church, or even the military, which is incorporated into the state itself, organized labor has a prominent role in the government party. This does not mean that it influences the decision-making process but, rather, that its relationship with the government, through the party, is formalized and visible.

A small percentage of unions in Mexico are independent of government control. In some cases their leadership has been able to obtain better benefits for members than have government-controlled unions. On the other hand, studies of independent unions in Mexico reveal that democratically elected leaders typically do not better represent the demands of the rank and file than the designated leaders in government-controlled unions.

The government has used unions to prevent the mobilization of large-scale opposition. "The government treats labor as a firm parent would a teenager. When it needs support in family crises and labor quickly provides it, it rewards the action. But when labor strays away from the family fold, it is scolded in a variety of ways. The government, not organized labor, controls the relationship."[34] The most common technique used by the firm hand of government is to promote new unions and leaders to keep established unions in line. Organized labor has failed to achieve political and economic influence in Mexico because of its subordination and because it represents only a small proportion of all workers. Unions have also remained weak and dependent on government because most of their members do not pay dues; unions actually receive subsidies from the government.[35]

One of the most important groups in Mexico, the largest unit within the union of government workers, is the National Teachers Union (SNTE). A recent study of it provides interesting data on its techniques for conveying demands to the private sector and the government (see table 6-1). Mexican unions must convince the government that their demands are legitimate; otherwise, they cannot legally strike. Determinations of legality

Table 6-1 Means Used by Organized Labor in Mexico to Convey Demands: National Teachers Union

Union Means of Action	Percentage of Total
Partial strike	20.71
Meetings, marches, demonstrations	20.71
Strike	16.82
Indefinite strike	6.47
Parade in front of public buildings	5.82
Call for a demonstration	5.82
Rejection of salary increases	5.50
Denunciations in press conferences	5.12
Block streets and highways	4.20
Occupy educational institutions	3.55
Block access to offices	2.58
List of demands to authorities	1.94
Hunger strikes	.64

Source: Este País, June 1991, 32–34; based on an analysis of 237 newspaper articles, January–April 1991.

are made by conciliation and arbitration boards, on which the government representative holds the deciding vote. The strike is only one means of conveying demands; marches and demonstrations have become increasingly common. But strike threats bring pressure to bear both on government and on private-sector management.

Because teachers are federal employees, it is revealing to examine how the government itself responds to labor demands (see table 6-2). Rarely does it actually raise salaries; rather, it provides low-cost benefits—such as discounts for married teachers at government stores—or relies on dialogue or promises to resolve complaints. If it decides not to negotiate, it then takes a hard line, either refusing to discuss the issues or threatening to fire striking teachers.

Organized labor, although part of the government corporatist structure, did not favor the nomination of Carlos Salinas as the PRI presidential candidate. In fact, the leadership of the powerful petroleum workers union encouraged its members to support the opposition. Only a short time after his inauguration Salinas had the head of the union arrested and charged with a number of criminal violations. Salinas also engineered a change in the leadership of the teachers union. Critics want him to eliminate the cozy relationship between organized labor and government to bring about economic and political liberalization. Instead, Salinas has established control over recalcitrant union leaders.

In terms of impact on government policy, labor has little to say. In the past decade real wages have declined radically, and labor has not been able to obtain increases even equal to inflation. In the only studies of actual policy decisions in which labor and the private sector have a part, the private sector comes out on top.[36] Still, the private sector believes that labor has the greater say in economic policy than business does, primarily

Table 6-2 Government Responses to Organized Labor Demands: National Teachers Union

Government Response to Union Demands	Percentage of Total
Married teachers will receive discounts	42.02
Dialogue	14.49
Shut off dialogue	13.04
Promise to resolve issues	8.69
No funds available	8.69
Will fire teachers who miss three days	7.24
Offer a small salary increase	2.89
Reject violent actions	1.44
Reject actions	1.44

Source: Este Pais (June, 1991): 32.

because labor is represented formally in the party structure and business is not. This inaccurate perception suggests the psychological and symbolic importance of such formal standing and of government *rhetoric*, which favors the interests of the working class.

Intellectuals

The intellectual community has an amorphous relationship with the government. Some of its formal organizations are patronized by the state, others are independent. None speaks for the intellectual community, but they do provide some public prestige. The most salient quasi-governmental organization is the National College, whose members are prominent in all fields, including law, sciences, humanities, social sciences, and fine arts.

The relationship of the intellectual community to the state is much more a product of the relationship between the government and intellectual employment than between the government and intellectuals' organizations. Three sectors of the economy employ the vast majority of intellectuals: government, academia, and publishing. Unlike intellectuals in the United States, Latin American intellectuals, Mexican intellectuals among them, have a long history of employment in public life, either in a federal bureaucracy, especially the secretariats of foreign affairs and education, or in various political posts as governors, party leaders, and cabinet members.[37]

The lack of employment opportunities in Mexico has encouraged intellectuals to work for the government. This means that the government does not have to incorporate intellectuals formally into institutional relationships with the state because the majority have been state employees since the 1920s. Many intellectuals, desirous of maintaining greater autonomy, have sought employment in the most prestigious universities, especially those in Mexico City. They occupy teaching and administrative positions at the National Autonomous University and at the Colegio de México. Intellectuals have advantageous ties to the government because many were classmates of future politicians and others have been their teachers. Politicians often identify prominent intellectuals as having been their most influential professors.

If intellectuals influence societal ideas, they do so through the written medium. Intellectuals in Mexico, as in the United States and other countries, establish magazines to circulate their views. Magazines dedicated to particular schools of thought are typically the product of a group of individuals who share certain ideological principles. One Mexican has described the phenomenon.

> There are some good papers and excellent magazines here, but each one tends to be controlled by some group or interest. All of these, such as *Excélsior,* or the publications of the Colegio de México and the Fondo de Cultura Económica, are publications of elite groups. It is very difficult for a person who writes to publish in them if [he or she does] not belong to the group in control of that publication. . . . These groups exist because most intellectuals are receptive to ideas paralleling their own preferences. . . . Actually there are very few independent intellectuals in Mexico, or intellectuals who have not formed groups.[38]

Some of the more prominent contemporary intellectual groups in Mexico include those of Octavio Paz, who contributes to his journal *Vuelta,* and of Héctor Aguilar Camín, who directs the popular monthly *Nexos.* Other groups are associated with newspapers, and many intellectuals earn a portion of their income contributing essays to editorial pages.

The government's relationship to the intellectual community is also reflected in its attitude toward the media and censorship.[39] Although freedom of speech obtains throughout Mexico, freedom of the press depends upon the medium. Radio and television programming, which intellectuals rarely use, comes under strict supervision. Book publishing is very open, although a bit of censorship takes place. Typically, publishers engage in self-censorship. Some notable cases exist of government pressure on publications that carry criticism of its workings. One such publication, the daily newspaper *Uno más uno,* lost its independence in a move engineered by the Salinas government; it has since followed a pro-Salinas line. In other situations the government has threatened to cancel or has actually canceled advertising in offending publications. Most leading publications depend on 30 to 40 percent of their revenues from government advertising; this indirect subsidy encourages caution in beneficiaries.[40]

The dependence of intellectuals and the institutions that employ them on the largesse of the state, affects their relationship to the government. Some intellectuals have successfully pursued independent careers as economic opportunities have expanded, which practice may actually be on the rise. Even so, the intellectual community relies heavily on government to support its activities and to recognize its merits, thus legitimizing government in the eyes of the educated Mexican.

VOICES OF DISSENT

Whether or not Mexico's political model is a more or less orthodox example of a corporatist structure, the government has never successfully incorporated all potentially influential groups into its fold, nor all members of the groups just discussed. Indeed, many individuals who have come to oppose the government politically have been its former supporters. These include both intellectuals and political opposition leaders.

Mexico has allowed dissent but successfully controlled its level and tone for some time. The government has a structural advantage in terms of continuity of leadership and the dominance of its machine, the PRI, over the voting process. In an underdeveloped economy, the state's economic resources are overwhelming, and in Mexico those resources have been used to disarm and co-opt dissidents, be they peasant leaders, lawyers, labor organizers, or intellectuals. Co-optation is the process by which the government incorporates an individual or group into its ranks. Groups find it difficult to counter government influence over their leaders. Few individuals can resist the attraction of political power or money, and the government often rewards cooperation with prestigious posts. Some indi-

---

Cooptation: the process by which the government successfully incorporates an individual or group into its ranks.

---

viduals accept the posts for financial reasons; others because of the possibility of working within rather than outside the system.

On the whole, the government has dealt well with contending groups, maneuvering them against one another when it believed that was necessary or creating intragroup competition to diminish the strength of a single recalcitrant leader or organization. The attitudes of each administration toward various groups and individual leaders has varied. President Salinas is altering overall government-group relations, giving greater attention and consequently prestige to business, the military, and the Church, and less attention to labor. Although changing somewhat in the 1990s, the relationships are unlikely to become radically different.

CONCLUSION

From the leadership's viewpoint, Mexico has developed a successful corporatist structure for engaging and controlling the society's most important interest groups. The corporatist system has never been comprehensive or complete but has channeled many political demands through quasi-governmental institutions. It is not an ideal corporatist system but is nevertheless essential to the workings of the political process.

Various groups in Mexico, including the military, the Church, business, labor, and intellectuals, have maintained somewhat different relations with the government, depending on the legal and institutional role given to them by society. Interestingly, whether their relationship is established and visible or more autonomous and independent, the informal channels their leaders use carry more weight in the decision-making process than do the formal channels. The political system has used and abused interest-group institutions to mobilize the rank and file for their own purposes rather than, for the most part, to hear group demands.

The groups having the most institutionalized relationship with the government through their incorporation in the party structure, have exercised the least influence on the whole over the decision-making process. Groups excluded from the party such as business, the Church, and the military, have influenced the decision-making process more heavily. Of the major interest groups that exist in most Western polities, business has had the most influence on Mexican government policies, primarily in the area of economics. The relationship between business and the government has been symbiotic, benefiting both.

The state itself has often pursued its own policies, not in response to the demands or pressures from any particular group, but because of self-interest or its interpretation of societal interests.[41] In this sense, the state has been an actor in the decision-making process. It has the greatest potential for influencing the outcome of policy-making because it operates in a semiauthoritarian environment, and it mediates among the more traditional, competing interests. The importance of various groups and the weight of their demands on the government are undergoing change in the 1990s.

NOTES

1. For the Mexican version, see Ruth Spalding, "The Mexican Variant of Corporatism," *Comparative Political Studies* 14 (July 1981): 139–61.

2. Nora Hamilton, *The Limits of State Autonomy: Post Revolutionary Mexico* (Princeton: Princeton University Press, 1982), describes this early pattern. Presidents Adolfo López Mateos (1958–1964) and Luis Echeverría (1970–1976) also responded more strongly to working-class interests.

3. For an interpretation that identifies cracks in the corporatist edifice in the mid-1980s, see Howard J. Wiarda, "Mexico: The Unravelling of a Corporatist Regime?" *Journal of Inter-American Studies and World Affairs* 30 (Winter 1988–1989), 1–28.

4. Luis Rubio, "Economic Reform and Political Liberalization," in *The Politics of Economic Liberalization in Mexico,* ed. Riordan Roett (Boulder, Colo.: Lynne Rienner, forthcoming 1993).

5. James Sánchez Susarrey, "Corporativismo o democracia?" *Vuelta* 12 (March 1988): 12–19.

6. Robert J. Shafer, *Mexican Business Organizations* (Syracuse: Syracuse University Press, 1973), explains these requirements in some detail.

7. Judith A. Teichman, *Policymaking in Mexico: From Boom to Crisis* (Boston: Allen & Unwin, 1988).

8. Miguel Basáñez, *La lucha por la hegemonía en México, 1968–1990,* 8th ed. (México: Siglo XXI, 1990), 35ff.

9. Abraham Lowenthal and J. Samuel Fitch, *Armies and Politics in Latin America,* rev. ed. (New York: Holmes & Meier, 1986), 4ff.

10. Franklin D. Margiotta, "Civilian Control and the Mexican Military: Changing Patterns of Political Influence," in *Civilian Control of the Military: Theories and Cases from Developing Countries,* ed. Claude E. Welch, Jr. (Albany: State University of New York Press 1976).

11. Gordon C. Schloming, "Civil-Military Relations in Mexico, 1910–1940: A Case Study" (Ph.D. diss., Columbia University, 1974), 297.

12. Jorge Lozoya, *El ejército mexicano (1911–1965)* (México: Colegio de México, 1970), 64.

13. James W. Wilkie, *The Mexican Revolution: Federal Expenditure and Social Change since 1910,* 2d ed. (Berkeley: University of California Press, 1970), 100–106.

14. Merilee Grindle, "Civil-Military Relations and Budget Politics in Latin America," *Armed Forces and Society* 13 (Winter 1987): 255–75.

15. Michael J. Dziedzic, "Mexico's Converging Challenges: Problems, Prospects, and Implications" (unpublished manuscript, United States Air Force Academy, April 1989), 34.

16. See my *Generals in the Palacio: The Military in Modern Mexico* (New York: Oxford University Press, 1992), for an analysis of these roles.

17. Phyllis Greene Walker, "The Modern Mexican Military: Political Influence and Institutional Interests" (M.A. thesis, American University, 1987), 76.

18. Soledad Loaeza, "La Iglesia católica mexicana y el reformismo autoritario," *Foro Internacional* 25 (October–December 1984), 142.

19. The relationship is outlined in Karl Schmitt, "Church and State in Mexico: A Corporatist Relationship," *Americas* 40 (January 1984): 349–76.

20. For an excellent example, see Matt Moffet, "In Catholic Mexico, A Priest's Power Is Limited to Prayer," *Wall Street Journal,* December 6, 1989.

21. For examples of the potential conflict politically, see Claude Pomerlau, "The Changing Church in Mexico and Its Challenge to the State," *Review of Politics* 43 (October 1981): 540–59.

22. Interview with Abelardo Alvarado Alcantara, auxiliary bishop of Mexico City, June 2, 1987.

23. Dennis M. Hanratty, "The Church," in *Prospects for Democracy in Mexico,* ed. George Grayson (New Brunswick, N.J.: Transaction, 1990), 118.

24. Interviews with former presidents José López Portillo and Miguel de la Madrid, Mexico City, summer 1990.

25. *Proceso,* October 14, 1991, 18–21.

26. The best analysis in English of these industrial groups is Dale Story, *Industry, the State, and Public Policy in Mexico* (Austin: University of Texas Press, 1986).

27. Roderic A. Camp, *Entrepreneurs and Politics in Twentieth Century Mexico* (New York: Oxford University Press, 1989), 141–42.

28. John Womack, "The Spoils of the Mexican Revolution," *Foreign Affairs* 48 (July 1970): 677–87.

29. Saúl Escobar Toledo, "Rifts in the Mexican Power Elite, 1976–1986," in *Government and Private Sector in Contemporary Mexico,* ed. Sylvia Maxfield (La Jolla: U.S.-Mexican Studies Center, UCSD, 1987), 79.

30. Graciela Guadarrama S., "Entrepreneurs and Politics: Businessmen in Electoral Contests in Sonora and Nuevo León," in *Electoral Patterns and Perspectives in Mexico,* ed. Arturo Alvarado (La Jolla: U.S.-Mexico Studies Center, UCSD, 1987), 83ff.

31. Howard Handleman, "The Politics of Labor Protest in Mexico: Two Case Studies," *Journal of Inter-American Studies and World Affairs* 18 (August 1976): 267–94.

32. George W. Grayson, *The Mexican Labor Machine: Power, Politics, and Patronage,* Significant Issues Series, vol. 19, no. 3, (Washington, D.C.: CSIS, 1989), 12.

33. For background, see Kevin J. Middlebrook, "State-Labor Relations in Mexico: The Changing Economic and Political Context," in *Unions, Workers, and the State in Mexico,* ed. Middlebrook (La Jolla: United States-Mexican Studies Center, UCSD, 1991), 1–26.

34. Roderic A. Camp, "Organized Labor and the Mexican State: A Symbiotic Relationship?" *Mexican Forum* 4 (October 1984): 4.

35. Kevin Middlebrook, "The Political Economy of State-Labor Relations in Mexico" (Paper presented at the National Latin American Studies Association, Washington, D.C., March 1982), and "The Sounds of Silence: Organised Labour's Response to Economic Crisis in Mexico," *Journal of Latin American Studies* 21 (May 1989): 195–220.

36. Susan K. Purcell, *The Mexican Profit-Sharing Decision: Politics in an Authoritarian Regime* (Berkeley: University of California Press, 1975).

37. Fred P. Ellison, "The Writer," in *Continuity and Change in Latin America,* ed. John J. Johnson (Stanford: Stanford University Press, 1964), 84.

38. Roderic A. Camp, *Intellectuals and the State in Twentieth-Century Mexico* (Austin: University of Texas Press, 1985), 131.

39. Albert L. Hester and Richard R. Cole, eds., *Mass Communications in Mexico* (Brookings, S.D.: Association for Education in Journalism, 1975).

40. Marvin Alisky, "Government Mechanism of Mass Media Control" (Paper presented at the Southeast Council of Latin American Studies, Tampa, April 1979).

41. For support of this view, see Rose J. Spalding, "State Power and Its Limits: Corporatism in Mexico," *Comparative Political Studies* 14 (July 1981): 139–64.

7

Who Governs? The Structure of Decision Making

> Once the Mexican president and his advisers are in agreement
> regarding the wisdom of making the decision, the president
> publicly associates himself with it by making a formal an-
> nouncement or an executive-sponsored legislative proposal, or
> both. All important decisions are formally initiated by the pres-
> ident, and the president both claims and receives full credit for
> the decision, whether or not the idea for the decision was
> originally his. Because of the patrimonial nature of staff ar-
> rangements, all individuals who participate in the decision-
> making process supposedly do so at the president's will and
> serve in the capacity of his subordinates. In return for receiving
> the delegated power to serve, they attribute all credit for their
> accomplishments to their patrimonial leader, the president.
>
> SUSAN K. PURCELL, *The Mexican Profit-Sharing Decision*

Every political system devises a set of structures and institutions to facilitate political decision making. Studies of decision making reveal that there
are a number of interrelated steps in the process. The steps begin with a
problem requiring a political solution, and pass through a series of institutions in which the problem is ignored or resolved, often legislatively. Some
institutions primarily channel demands from society through the political
system. Other institutions contribute to the selection and election of political leadership. Still others carry out the solutions proposed by the political
system.

Each political model performs the steps in decision making differently, although many models have certain similarities. For example, in the
United States, the legislative branch plays a critical role in the formulation
of laws and as a focus of interest group activity. In the United Kingdom
although Parliament plays a critical role in approving legislation, most of

131

its formulation and lobbying is done through the executive branch. The cabinet, however, is a product of the legislative branch, that is, its members are members of Parliament; thus election to Parliament determines who will make many government decisions.

Mexico, as has been suggested earlier, evolved a political system that formally resembles that of the United States but centralizes much greater authority in the executive branch. The powers of the executive branch combined with the dominance of a leadership group represented by a single party—the PRI and its antecedents—has led to a government dominated by the executive, largely in the person of the president. Which institutions are the most salient, and what functions do they perform?

THE EXECUTIVE BRANCH

The seat of the Mexican government is Mexico City, in the Federal District, a jurisdiction with certain similarities to the District of Columbia in the United States. Mexico City, however, unlike Washington, D.C., combines the qualities of New York City, Chicago, and Los Angeles, for Mexico's political capital is also its intellectual and economic capital.

The executive branch of the government houses two types of agencies: those that have counterparts in most First and Third World countries, such as secretariats of foreign relations and national defense, and others that are idiosyncratically Mexican, sometimes called decentralized or parastatal agencies, somewhat analogous to the Tennessee Valley Authority in the United States. Parastatal agencies are a product of Mexican nationalism, Mexicanization, and state expansion from the 1940s through the 1980s, culminating in the nationalization of private, domestically owned banks in 1982.[1]

The preeminent parastatal agency in Mexico, recognized internationally, is Petroleos Mexicanos (Pemex), the national petroleum company. Pemex was born when President Lázaro Cárdenas nationalized foreign-owned petroleum companies in 1938.[2] Since then the government has controlled the development of petroleum resources, including exploration and drilling, and the domestic retailing of petroleum products. Because of the vast Mexican oil reserves and their rapid exploitation in the 1970s and 1980s, Pemex became Mexico's number-one company. Its sales at their apex accounted for more than three-quarters of export revenues.

Among the fifty leading firms (excluding banks) in Mexico during the 1980s, a fourth were government owned. Other important government entities include the Federal Electric Commission, which develops and distributes electricity; the National Bank of Foreign Commerce, designed to promote trade; the National Company of Public Commodities (Conasupo), a distributor of basic foodstuffs to low-income Mexicans; Sidermex, the basic-steel producer; the National Finance Bank (Nacional Financiera), a developmental bank; and many other companies in utilities, communications, transportation, minerals, fertilizers, and so on. The agencies mentioned above have semicabinet status, and the president announces his appointees to them simultaneously with those of formal cabinet members.

The formal cabinet comprises twenty-one agencies: Attorney General of the Republic; Attorney General of Justice for the Federal District; Secretariat of the Comptroller General; Secretariat of Fishing; Department of the Federal District; Secretariat of Agrarian Reform; Secretariat of Tourism; Mexican Institute of Social Security; Secretariat of Agriculture and Hydraulic Resources; Secretariat of Communications and Transportation; Secretariat of Foreign Relations; Secretariat of Government; Secretariat of Energy, Mines and Government Industries; Secretariat of Health and Welfare; Secretariat of Commerce and Industrial Development; Secretariat of Labor and Social Welfare; Secretariat of National Defense; Secretariat of the Navy; Secretariat of Social Development; Secretariat of Public Education; and Secretariat of the Treasury and Public Credit.

An examination of the major agencies suggests some interesting aspects of Mexican policy issues, and the importance of specific economic problems. For example, the historic impact of agrarian issues and agrarian reform after the Revolution can be seen in the fact that *two* cabinet-level agencies are devoted to agriculture, one specifically to agrarian reform, and until recently, hydraulic resources were the purview of a separate agency. Nevertheless, it would be misleading to say that any president since Lázaro Cárdenas has given priority to agrarian issues. In fact, the desire of the Salinas administration to eliminate land-tenure problems generated by village-held land titles (*ejidos*) may mean the disappearance of the Secretariat of Agrarian Reform in the immediate future. A second agency of special importance is Tourism, which has had departmental status since 1959, indicative of the industry's impact on the economy. The most recently reconstituted secretariat is that of Social Development, in response to its political and economic importance. Something that probably would seem unusual to most Americans is the separate attorney general and department for the Federal District. The Federal District, which now accounts for nearly a fifth of the total population, has no direct, elected executive leadership. Therefore, district residents are governed, in effect, by a presidential appointee. This post has come to be of increasing consequence as the city and entire metropolitan area have rapidly expanded in scope and become more densely settled. In response to residents' complaints that they had no direct representation, the city now has a forty-member Federal District assembly, but—as in the case of state and national structures—it is the executive branch, through the Department of the Federal District, that has political power.[3]

The agencies of greatest standing in the executive branch are typically those with long histories. In the 1920s, 1930s, and 1940s, the Secretariat of National Defense carried far more weight than it does today, not because of its impact on day-to-day policies but because it often was the source of presidential leadership. Given the centralization of power in the hands of the president and, as we have seen, the importance of individual, federal bureaucratic agencies as sources of political recruitment, some relationship exists between decision-making influence and the degree to which individual agencies are the source of high-level personnel. In the 1950s and 1960s, the Secretariat of Government, an agency devoted to internal political affairs, replaced the Secretariat of National Defense as a source of presidential leadership, and as a major voice in policy decisions.[4]

Despite the roles played by the Secretariat of Defense and the Secretariat of Government, the Secretariat of the Treasury wielded considerable influence, and its head received much attention in each cabinet. President Cárdenas enhanced treasury's authority, and its leader by authorizing him to act as an arbiter in the allocation of funds to other agencies and to state governors in connection with the federal revenue sharing program.[5] Thus, other than the president, the treasury secretary became the key figure in the distribution of economic resources, as well as in the determination of the direction of financial policy.

Economic agencies in the government gained substance with the onset of hard times. By the 1980s, the Secretariat of Programming and Budgeting (combined with treasury in 1992), the Secretariat of the Treasury, and the Bank of Mexico (the federal reserve bank) became the troika in setting economic policy.[6] To streamline cabinet coordination and facilitate policy-making, Miguel de la Madrid organized subcabinet groups along policy lines, including an economic cabinet. These groups have been more active under Salinas, and he has added another category, national security, giving it heightened visibility. It includes the Secretariats of Government, Foreign Relations, National Defense, and the Attorney General of the Republic.

Groups in Mexican society who want some part in national policy decisions must make their concerns and interests known to the executive branch at the highest possible level. Yet, as Daniel Levy argues, this is difficult to accomplish:

> As most important legislation is initiated and carried through to approval by the president, hardly any opportunity exists for effective interaction between citizens and their representatives during the lawmaking process. However, groups and individuals may occasionally influence the way in which laws and policies are actually implemented. A common element of day-to-day politics in Mexico is the presentation of demands to local and state governments, to departments of the federal bureaucracy, and even directly to the president.[7]

The cabinet secretary is the key figure in initiating policy proposals, and his staff thoroughly studies the issues and collects information relevant to the formulation of policy. He may be responding to a presidential request or pursuing matters associated with his agency's mandate under broad guidelines outlined to him by the president and the presidential advisers.[8] The individuals who have access to the president himself are even more successful in influencing decisions than are persons whose highest contacts are cabinet figures.

Because the decision-making structure is so hierarchical and the president exercises so much influence, or is expected to exercise authority over the system, considerable pressure is put on channels of access to the presidency. The president's private secretary, who in effect functions as a chief of staff, and whose position is essentially a cabinet-level appointment, has the complete confidence of the president. Because he acts as a gatekeeper in denying or acceding to requests to see the president, he performs a crucial role in the decision-making process.

Salinas has given a special emphasis to two positions in his administration, positions that reflect the nature of the decision-making process. To coordinate the cabinet and keep closer control over policy initiatives, the president appointed a coordinator of the technical cabinet subgroups who reports directly to him. The coordinator is a naturalized citizen, José Córdoba, who because of his foreign birth is constitutionally ineligible to be either a cabinet secretary or president. The fact that Córdoba is not a native Mexican has produced certain tensions among those who are close to the president; on the other hand, the limits on his future political ambitions give the president a certain confidence in his aide's motivations.

Part of any decision-making process involves informing the public about policy decisions. Salinas understands public opinion better than any recent Mexican president. Consequently, he has given much thought to the position of head of social communication for the presidency, the Mexican version of the U.S. presidential press secretary. But more than coordinating presidential press conferences, a rarity in Mexico, the individual attempts to shape media coverage of presidential actions, policy initiatives, and policy outcomes. A positive "spin" is the desideratum not only domestically but in all other quarters as well.[9]

The few studies that have been carried out of Mexican decision making have attempted to classify the role of the presidency in the decision-making process.[10] Although there is no question that decision making is centralized, and that the president personally has greater influence over the outcome of policies than does the U.S. president because more than 90 percent of legislation comes from his office, he cannot in most cases arbitrarily make a policy decision—nor is it likely he would desire to do so. The worst fears of critics of the Mexican semiauthoritarian decision-making process were borne out when President José López Portillo announced without warning the nationalization of the banks in 1982. The circumstances surrounding the decision have been well documented, and according to the few individuals López Portillo consulted, he did not consider the views of any of the groups that would be affected.[11] The fact that a single political actor, in consultation with two or three individuals, could make a decision that would have major reverberations throughout the economy and bring relations between the private sector and the state to a breaking point demonstrates the dangers inherent in centralized power.[12]

Typically, however, presidents do not operate in solitary splendor, but their consultations tend to be more private and hidden from public view than in the United States, where lobbying goes on in front of the scenes as well as behind them. One of the characteristics of the decision-making process is that often it is the executive branch itself that takes the initiative in regard to affected parties rather than vice versa. In other words, the role of interest groups is often reactive, not proactive.

THE LEGISLATIVE BRANCH

Mexico's national legislature is bicameral: the Chamber of Deputies and the Senate. Deputies are elected on the basis of roughly equally populated districts, of which there are three hundred. In 1970 one hundred seats were added for deputies selected from party lists based on the proportion of the votes cast for the parties. The purpose of the increment was to increase opposition representation, given the overwhelming dominance of the PRI in the regular legislative seats. In the reforms of the 1980s, another hundred seats were added; now three hundred deputies represent districts and two hundred represent parties. The Senate, which has fewer powers than the Chamber of Deputies, has two senators from each state and the Federal District, a total of sixty-four. Senators are elected for six-year terms, until 1988 coterminous with the presidential term. Beginning in 1991 half the Senate's membership is elected every three years.

The Chamber of Deputies and the Senate each have numerous committees, some with names like those in the U.S. Congress. But because deputies and senators cannot be reelected to consecutive terms, seniority does not exist, at least regarding committees, for all members are new to a particular legislature. Some critics argue that one means of enhancing legislative powers in Mexico is to allow consecutive reelection, which would permit members to develop stronger ties with their constituencies. Interestingly, many Americans would like to see a limit set on congressional terms.

An examination of the committee structure reveals that in many cases congressional leaders attempt to place individuals on committees with some relevance for their expertise and/or interest. For example, in the past, military officers on leave or retired were appointed to the National Defense Committee, or legislators representing the peasant unions were given assignments on the committees dealing with agriculture.

The legislative branch has long been controlled by the PRI, whose members have accounted for more than 90 percent of the district seats in the Chamber of Deputies, and until 1988, all Senate seats (see table 7-1). The president appoints a congressional leader (equivalent to the majority

Table 7-1 Representation in the Legislative Branch, Mexico, 1991–1994

| Party[a] | Deputies | | | Senators |
	District Seats	Party Seats	Total	
PRI	290	31	321	61
PAN	10	80	90	1
PRD	0	40	40	2
PFCRN	0	23	23	0
PARM	0	14	14	0
PPS	0	12	12	0
Total	300	200	500	64

Source: The Other Side of Mexico, no. 22 (July–August 1991): 4.

[a] PRI = Institutional Revolutionary Party; PAN = National Action Party; PRD = Democratic Revolutionary Party; PFCRN = Cardenista Front for National Reconstruction Party; PARM = Authentic Party of the Mexican Revolution; PPS = Popular Socialist Party. The other parties, Mexican Labor, Ecologist, Democratic, and Revolutionary Workers, did not win 1.5 percent of the vote, thereby losing their national registration.

leader in the U.S. Congress), who heads all of the state delegations. Each state's delegation in the Chamber is usually headed by an individual who has served before or by a rising star who is given the post for the first time. Many deputies from the PRI complain that decisions are made in an authoritarian fashion by the leadership, and that as individuals, they play a minor role.[13] Opposition members serve on the various committees as well. The Senate also has a leader, and the senior senators from each state form the internal governing body.

Earlier discussion indicated that the legislative branch had little to say in the decision-making process, unlike the U.S. Congress. The reason for this is that each legislator who is a member of the government party is beholden to the political leadership, and indirectly the president, for his or her position. If such a legislator wants to pursue a public career, he must follow presidential directives. What, then, does Congress do? It serves in several capacities. One, it examines presidential legislation initiatives and makes recommendations to the executive branch for alterations. Although theoretically the legislature can reject a presidential initiative, most presidential legislation is approved. Before the presence of opposition parties in Congress, it was approved overwhelmingly. Although they can be ignored by the executive branch, the legislative leadership reports opposition legislators' criticisms and suggestions.

Beginning with the administration of José López Portillo, the legislative branch took on a practice common to the British Parliament: cabinet secretaries are required to come before the Chamber yearly and report on their various activities. Again, although the Chamber is powerless in reality to alter cabinet decisions or to withhold resources, discussions of the reports and opposition pronouncements are covered in the media. The Chamber is a forum where public policy can be debated, even if that forum is limited in scope.

The Senate does not pass on legislation, but approves or disapproves of certain executive branch appointments—just as the U.S. Senate does with presidential appointments. Because senators are elected under the same conditions as PRI deputies, they are not likely to reject a presidential appointment. There have been some cases, however, when this has occurred, most notably in connection with military promotions. All career military officers above the rank of colonel are promoted by the president subject to the pleasure of the Senate. In the early 1950s the Senate actually rejected an abuse of presidential authority involving promoting officers who had not met the required time in grade according to military law.[14] Recent presidents have not violated their authority in this regard.[15]

The legislative branch serves to legitimize executive legislation. One of the potential consequences of the 1988 elections, when opposition parties obtained the largest representation ever in the Chamber of Deputies and the Senate, was that the government lost its ability to amend the Constitution; the PRI did not have two-thirds of the seats in the lower chamber, the number necessary to do so. Mexican presidents and the executive branch have used constitutional amendments to give major, controversial legislation an extra measure of legitimacy. The opposition's gains in 1988 prevented the government from using this technique without first achieving a coalition. Even in general terms, this became true. As Wayne Cornelius, Judith Gentleman, and Peter Smith point out,

> Salinas's dealings with the Congress undoubtedly were made more complex by the election results, which created, for the first time in Mexican history, the conditions for vigorous multiparty competition within that branch of government and between Congress and the president. Salinas will be the first post revolutionary president who has had to genuinely negotiate with opposition forces in Congress on a wide range of issues.[16]

The election results of August 1991, however, reversed the situation, once again giving the PRI overwhelming control of the chamber.

The legislative branch is also a training ground for future political leaders and an important source of political patronage. It has been used to reward people prominent in quasi-governmental interest groups and among the labor, peasant, and popular, professional sectors.[17] Although professional people predominate among the legislators, peasant and labor leaders, as well as women, who might not obtain higher political office in the executive branch, are well represented. Even more important, the legislative branch provides upward mobility to a different type of politician: individuals who are more likely to have come from a working-class background, from the provinces (because of the district representation), from electoral careers, and with less formal education (see table 7-2). Women too, as in many European countries, are best represented in this branch of government. In short, greater percentages of individuals who are excluded from executive-branch careers, even at the departmental level, can find places in the legislative branch. The fact that some channels are open for these kinds of Mexicans, who in many background characteristics correspond more closely to the population in general, is important for social mobility and leadership fluidity.

The legislative branch also serves as a training ground for political skills. Among government institutions, opposition leaders and parties are only represented in the legislative branch. Their electoral wins in 1988 and 1989 temporarily increased their political influence, forcing government

Table 7-2 Legislators and Executive-Branch Officials, 1989–1991

Background Variable	Legislators (%)	Executive Branch Officials (%)
Gender		
Female	12	5
Education		
Preparatory or less	21	2
Career Experience		
Political parties	78	41
Unions	54	16
Elective posts	99	7
Parents' Occupation		
Peasant	7	1
Laborer	3	1
Birthplace		
Federal District	13	51

Source: *Diccionario biográfico del gobierno mexicano* (Mexico: Presidencia de la República, 1989); based on 1,113 officials and 560 legislators.

leadership to compromise on several policy issues related to electoral reform.[18] Negotiating skills will be more and more valued in the decision-making process if opposition parties continue their progress in vote getting. Most officials in the executive branch have little or no experience in such skills, hence persons whose careers have brought them through the legislative bodies are likely to be in greater demand in the future. For example, when the president removed the newly elected governor of San Luis Postosí in 1991, he appointed his former mentor and his majority leader in the 1988–1991 legislature, Gonzalo Martínez Corbala, interim governor, a difficult assignment requiring superior political skills.

THE JUDICIAL BRANCH

A major principle in the U.S. government structure is the balance of power. The founding fathers were concerned that the executive branch might take on dictatorial aspects and hence sought to apportion power among the executive, legislative, and judicial branches in such fashion that none could assume paramountcy. The Mexican judicial system is structurally patterned after that found in the United States. It has local, state, and national levels, the last comprising a court of appeals and a supreme court.

A judicial branch influences the decision-making process when it is independent of legislative and executive authority, and when it can legislate through judicial rulings. The U.S. Supreme Court can declare a law unconstitutional, after which Congress can devise other legislation to achieve its goal if it so wishes. U.S. courts hand down rulings that bear on future cases and also on legislation regarding the issues involved.

Legislating through judicial precedent is not a viable procedure in Mexico. For the Supreme Court to establish a binding precedent, it has to reach identical conclusions about precisely the same issues repeatedly. This rarely, if ever, occurs. Although the Supreme Court has some independence, justices do not sit for life and their appointments are political. Although they are generally praised for their integrity, they do not venture into political issues. The high bench typically rules on appeals of individuals, not on matters of constitutionality.

The lower levels of the legal system are tainted by corruption and outside political manipulation. An absence of consistency and integrity makes it difficult, if not impossible, for the average citizen to resort to the system to protect his or her rights. The criminal justice subsystem has incorporated the use of torture in obtaining confessions.[19] These circumstances combine to create a lack of respect for the law, a crucial element in a viable, legal system.

THE PRI

Many analysts of Mexican politics commonly refer to the government as the PRI or, frequently, the PRI government. The label implies that the PRI, which is the political party of the government, exercises policy-making authority over the system. Nothing could be further from the truth. As pointed out in the previous chapter, the PRI plays a significant role in institutionalizing corporatist structures and in the relationship between certain groups and the government. In fact, the PRI acts as a channel in decision making for the least influential groups. Its own leadership has little if any impact on the making of policy, as Dale Story suggests:

> The Party [the PRI] clearly does not control the reins of political decision-making, nor is it even a coequal to the state. Yet most national elites are at least Party members, and more significantly, the Party is a very critical institution serving the executive branch of government, in particular the office of the presidency. Especially with national elites becoming so technocratic, the PRI provides the president with the necessary political legiti-

macy, the symbolic aura of the Revolution, and the machinery for running campaigns, winning elections, and maintaining contacts with the masses.[20]

As an institution, the PRI does not have policy influence over the members of the legislative or executive branches. Although its role is very visible in the legislative branch because its members are the leaders of both chambers, they do not report to the Party leadership. Even if they were to do so, the Party leadership is selected by the president, thus placing the Party under the thumb of the executive branch. The Party relies on the executive branch for financial support; generally, the Secretariat of Government allocates the funds. The support is difficult to measure because it involves more than money. The government, through its contacts, provides many other resources, such as lodging, transportation, and meals for those doing Party business. Individual candidates receive little direct financing from the Party, but it does pay for Party, as distinct from candidate, advertising, indirectly promoting the fortunes of the individual politician.[21]

The PRI, as a Party vehicle, does not even have much influence over executive branch officials, who are the most active in the decision-making process. Many of the officials have few formal ties to the Party; in fact, a number of top officials have never been members of any party. Credentials other than active Party experience are of greater value to an individual's career.

The PRI does not function autonomously. Its dependence on the government and on executive branch leadership effectively eliminates any direct influence it might have on the decision-making process, especially in connection with economic and social policy issues. Nevertheless, in terms of recent political reforms, officials who have made their careers within the PRI, particularly at state and local levels, have begun to express themselves as a viable interest group on executive-branch decisions affecting the strength and growth of the Party. The executive branch has demonstrated its superiority in the decision-making process in imposing solutions on Party problems. This is illustrated clearly in two races during the August 1991 elections. The president removed two PRI gubernatorial candidates—one before he took office in Guanajuato, the other soon after his inauguration in San Luis Potosí—after the PRI had claimed substantial victories in those states. The opposition had cried fraud, and in the latter state, marched on the national capital, but the PRI rank and file felt betrayed. Whether or not the PRI had won fairly, the president imposed his will on the party leadership, making clear their subordination to presidential authority.

The PRI, in spite of fraud, has shown its ability to adapt and survive. If the political leadership wishes to rely on the PRI to continue legitimizing its authority through the electoral process, as opposition strength increases, then the PRI bureaucracy will gain in influence in the political arena. Party officials who make their political careers in the Party bureaucracy and in elective office, will develop and express their own interests, as do officials in the federal bureaucracy, attempting to have a say in decisions that affect their institutional future as well as their political careers.

CONCLUSION

Decision making in Mexico is controlled through the executive branch, centralized in the person of the president. As economic problems have overshadowed all other issues, the influence of the economic cabinet has expanded. The decision-making process listens to demands more through informal internal channels than through formal public channels. As the analysis of interest groups in chapter 6 demonstrates, leaders from various sectors seek out individual decision makers in the executive branch, typically the cabinet secretary or, if they have access, the president.

The degree of centralization of decision-making power in the president and the executive branch flavors the whole governmental process. Not only does a president have a huge reservoir of political authority but most Mexicans expect, indeed, react positively to, his exercise of his powers. Salinas was praised for his decisiveness during the first three years of his administration, when he used his decision-making authority to rebuild lost confidence in the presidency.

The reliance on informal channels of influence favors certain groups over others. Business interests have been more successful than labor or peasants in having their point of view heard. The government does not stress listening to demands made through formal channels; rather it concerns itself strongly with how its policies are received, and its image. It attempts to use the media to promote its effectiveness as a decision maker, and its most notable success under Salinas has been the Solidarity Program, which has brought him personal support, and brought the PRI political support. A 1991 poll in *Este País* found that 62 percent of its respondents thought highly of the program, and four in five who would vote for the PRI were favorable toward Solidarity.[22] Yet that program, like other government initiatives resulted from a top-down approach to the allocation of resources rather than its reverse.

The powers exercised by the executive branch since the Revolution have left Mexico with weak legislative and judicial institutions. Not only did the state grow in size throughout most of this period, reversed only since 1988, but its power lay within the executive branch. Because ambitious politicians understand this, competition for careers in the executive is more intense than in the other branches. In fact, the imbalance has discouraged formation of an active, independent opposition, which contributes to the leadership's co-optive capability.

NOTES

1. For an analysis of how this sector has functioned, see the case study by William P. Glade, "Entrepreneurship in the State Sector: Conasupo of Mexico," in *Entrepreneurship in Cultural Context,* ed. Sidney Greenfield et al. (Albuquerque: University of New Mexico Press, 1979), 191–222.

2. For background, see George Grayson, *The Politics of Mexican Oil* (Pittsburgh: University of Pittsburgh Press, 1980); Edward J. Williams, *The Rebirth of the Mexican Petroleum Industry* (Lexington, Mass.: Heath, 1979).

3. For an excellent analysis of the historical evolution of the Federal District, and a discussion of the pros and cons for representation within this entity, see Peter M. Ward, "Government without Democracy in Mexico City: Defending the High Ground," in *Mexico's Alternative Political Future,* ed. Wayne A. Cornelius et al. (La Jolla: Center for U.S.-Mexican Studies, UCSD, 1989), 307–24.

4. Miguel Alemán, 1946–1952; Adolfo Ruiz Cortines, 1952–1958; Gustavo Díaz Ordaz, 1964–1970; and Luis Echeverría, 1970–1976, are presidents indicative of this changing institutional influence and the rise of civilian leadership; they came from the Secretariat of Government.

5. See the introduction by Antonio Carrillo Flores in Eduardo Suárez, *Comentarios y recuerdos, 1926–1946* (México: Porrúa, 1977).

6. For background on the rise of the Secretariat of Programming and Budgeting, see John J. Bailey's excellent "Presidency, Bureaucracy, and Administrative Reform in Mexico: The Secretariat of Programming and Budgeting," *Inter-American Economic Affairs* 34 (Summer 1980): 27–59.

7. Daniel Levy and Gabriel Székely, *Mexico, Paradoxes of Stability and Change,* 2d ed. (Boulder, Colo.: Westview Press, 1987), 49–50.

8. For case studies in education, hydraulic resources, and agricultural policy, see Guy Benveniste, *Bureaucracy and National Planning: A Sociological Case Study in Mexico* (New York: Praeger, 1970); Martin H. Greenberg, *Bureaucracy and Development: A Mexican Case Study* (Lexington, Mass.: Heath, 1970); Merilee S. Grindle, *Bureaucrats, Politicians, and Peasants in Mexico: A Case Study in Public Policy* (Berkeley: University of California Press, 1977).

9. For example, PRI headquarters now has a special section that provides information to foreign scholars; the government sends copies of its own newspaper, *El Nacional,* to various academics; and the embassy mails data on election results and speeches of party officials.

10. See, for example, Purcell, *The Mexican Profit-Sharing Decision,* 4; Roderic Ai Camp, *The Role of Economists in Policy-Making: A Comparative Case Study of Mexico and the United States* (Tucson: University of Arizona Press, 1977), 9; Judith A. Teichman, *Policymaking in Mexico: From Boom to Crisis* (Boston: Allen & Unwin, 1988).

11. Carlos Tello, *La nacionalización de la banca en México* (México: Siglo XXI, 1984).

12. Roderic Ai Camp, *Entrepreneurs and the State in Twentieth Century Mexico* (New York: Oxford University Press, 1989), 128–33.

13. The best description of how the committee system functions and the role of the Chamber of Deputies still is Rudolfo de la Garza, "The Mexican Chamber of Deputies and the Mexican Political System" (Ph.D. diss., University of Arizona, 1972).

14. Senado, *Diario de los Debates,* 1953, 5–6.

15. Personal interviews, Mexico City, 1990–1991.

16. Wayne A. Cornelius, Judith Gentleman, and Peter H. Smith, "Overview: The Dynamics of Political Change in Mexico," in *Mexico's Alternative Political Futures,* ed. Cornelius, Gentleman, and Smith (La Jolla: U.S.-Mexico Studies Center, UCSD, 1989), 25.

17. See Peter H. Smith, *Labyrinths of Power: Political Recruitment in Twentieth-Century Mexico* (Princeton: Princeton University Press, 1979), 217ff.

18. For background, see Robert A. Pastor, "Post-Revolutionary Mexico: The Salinas Opening," *Journal of Inter-American Studies and World Affairs* 32 (Fall 1990): 1–22.

19. Americas Watch, *Human Rights in Mexico: A Policy of Impunity* (New York: Human Rights Watch, 1990), 1; *Unceasing Abuses, Human Rights One Year After the Introduction of Reform* (New York: Human Rights Watch, 1991).

20. Dale Story, *The Mexican Ruling Party: Stability and Authority* (Stanford: Hoover Institution, 1986), 131–32.

21. Interview with a candidate for federal deputy, Mexico City district, 1985.

22. *Este País,* October 1991, 9–10.

8

Expanding Participation:
The Electoral Process

A cursory examination of the changes in the electoral laws and
practice since 1917 suggests that the Mexican system has been
gradually but perceptibly marching in a democratic direction.
Pre-World War II presidents such as Madero, Carranza,
Obregón, and Cárdenas received from 95 to 99 percent of the
total vote. And prior to 1939 no permanently organized opposition parties existed to challenge the PRI. . . . However, despite
the proliferation of parties in recent years and the lower vote
totals for the PRI, the PRI has never lost a presidential election
(and always carried at least two-thirds of the vote) nor a gubernatorial election and has lost very few elections for federal
deputies, senators, and scores of state and local offices.

DALE STORY, *The Mexican Ruling Party*

A little over a decade ago, most political analysis would have given little
space to elections and electoral politics in Mexico. Although elections have
been a feature of the political landscape since the time of Porfirio Díaz,
with the exception of Francisco Madero's election in 1911, they never
functioned as the crucial determinant of political leadership nor furnished a
policy mandate.

Beginning in the mid-1970s elections took on a new dimension. At
first, the uncharacteristic emphasis could be tied to the desire of some
establishment figures to strengthen the PRI's image and that of the political
system by promoting the opposition's fortunes. In other words, the government itself, through a series of electoral reforms, tried to stimulate the
opposition. It provided opposition parties with an incentive to challenge
the PRI's dominance by increasing their potential rewards but without
extending the possibility of real victory. The single-party dominance over

146

the system and election results was brought home when the National Action Party (PAN), frustrated by the futility of opposition, refused to nominate a candidate to run against José López in the 1976 presidential race.

ELECTORAL REFORMS

Some government strategists believed it was smart politics to increase opposition representation in the Chamber of Deputies through implementation of a plurinominal deputy system (deputies selected on the basis of their party's total national vote); others were committed to actual reforms. The latter, who hoped to democratize the elections, believed genuine competition would strengthen the political model and increase participation, a change for which they believed Mexicans were ready. One of the architects of these earlier reforms, Secretary of Government Jesús Reyes Heroles, introduced enabling legislation in 1977 under José López Portillo (1976–1982).[1]

The 1977 reforms, incorporated into the Federal Law of Political Organizations and Electoral Processes (LOPPE), altered several constitutional provisions. The law increased majority districts for federal deputies (similar to United States congressional districts) from approximately two hundred to exactly three hundred seats. It also specified that an additional one hundred seats were to be assigned to opposition parties in the Chamber of Deputies through a complex mathematical formula that allocated seats

---

Majority districts: legislative districts of roughly equal populations whose congressional representative wins the largest number of votes cast within the district.

---

proportional to each party's national vote totals. Under the party-deputy arrangement in effect from the 1964 through the 1976 national elections, opposition parties were allocated thirty to forty seats, also on the basis of each party's national vote totals. From the mid-1960s to the mid-1970s the opposition, combining party and majority deputies, averaged about 17 percent of the seats in the lower house but none in the Senate. The 1977 law, however, set aside *all* one hundred seats for this purpose, requiring them to be divided proportionately among the opposition parties. In effect, this meant that opposition parties, after the LOPPE went into effect, garnered approximately 26 percent of the seats in the lower house, and were guaranteed a minimum of one-quarter of all the seats.

The 1977 reforms, while giving some encouragement to opposition parties, providing them with more seats in the Chamber of Deputies, and allowing them greater access to the media during campaigns, lost their impetus after 1979, when Reyes Heroles left his post. Both the reforms in the early 1960s, which first introduced the party deputy system, and the 1977 electoral law, which created the plurinominal deputy system, stabilized opposition gains at a given level during the life of the legislation. In

> Proportional representation: a system for allocating legislative seats to parties on the basis of the national vote cast for all of the party's legislative candidates.

other words, opposition representation within the two periods (1964–1976 and 1979–1985) remained stable (see table 8-2) at approximately 17 and 26 percent respectively, suggesting at least on the basis of official tabulations that opposition parties experienced little growth.

When President de la Madrid took office in 1982, he seemed to indicate a new posture toward the opposition. Specifically, his administration tolerated intense electoral competition at the local level. Wayne Cornelius argued:

> Even more significantly, de la Madrid established a new policy regarding municipal elections: henceforth, municipal-level victories by opposition party candidates would be recognized, wherever they occurred. During the first ten months of his administration, the PRI conceded defeat in municipal elections held in seven major cities, including five state capitals and Ciudad Juárez, a large city on the U.S.-Mexican border. Virtually no electoral fraud was reported in these key municipal contests of 1983. As one high-ranking PAN official recalled, "It was like Switzerland up there. There was no interference in the voting, and the ballot count was absolutely clean.[2]

In 1986 de la Madrid introduced his own electoral law, which was to have significant consequences in the 1988 presidential elections, the first to test it. The 1986 electoral code included the following provisions:

1. The winning or majority party is never to obtain more than 70 percent of the seats in the lower chamber.
2. Three hundred deputies are to be elected by a relative majority based on individual congressional districts (similar to the United States).
3. The seats allotted to deputies on the basis of a proportional percentage of their total national vote are to be increased from 100 to 200, increasing the total number of seats from 400 to 500 (300 by district, and 200 by proportional representation).

4. Opposition parties may obtain 40 percent of the seats without winning a single majority district (200 of the 500 seats).
5. The party winning the greatest number of majority seats is to retain a simple majority in the entire deputies' chamber; that is, the party is to be allotted seats through the proportional representation system sufficient to obtain an absolute majority in the lower house.
6. Half the Senate is to be renewed triennially instead of the entire chamber every six years (the first change in senatorial elections since 1934).[3]

These reforms, compared to the two earlier laws, reduce the proportion of the majority party (the PRI) in the Chamber of Deputies, which ranged from 83 percent to 74 percent, to 70 percent or less. The 1986 law also reduced the overall importance of majority districts, which were generally won by the PRI (usually 95 percent or more, see table 8-1). Again, the law increased the opposition's presence in the Chamber of Deputies, but it was an increase *allocated by the government,* rather than an increase that the government permitted the opposition to earn.

It can be argued that economic and political conditions did more to boost opposition fortunes and to give elections greater political importance in Mexico than did internal reforms during this interim. As the economic crisis worsened in the early 1980s, the opposition on the state and local

Table 8-1 Percentage of Total Vote Won by Candidates for Congress by Major Party, 1961–1991

Election Year	Party[a]										
	PRI	PAN	PPS	PARM	PDM	PSUM	PST	PRT	PMT	PRD	PFCRN
1961	90.2	7.6	1.0	0.5	—	—	—	—	—	—	—
1964	86.3	11.5	1.4	0.7	—	—	—	—	—	—	—
1967	83.3	12.4	2.8	1.3	—	—	—	—	—	—	—
1970	80.1	13.9	1.4	0.8	—	—	—	—	—	—	—
1973	69.7	14.7	3.6	1.9	—	—	—	—	—	—	—
1976	80.1	8.5	3.0	2.5	—	—	—	—	—	—	—
1979	69.7	10.8	2.6	1.8	2.1	4.9	2.7	—	—	—	—
1982	69.3	17.5	1.9	1.4	2.2	4.4	1.8	1.3	—	—	—
1985	65.0	15.5	2.0	1.7	2.7	3.2	2.5	1.3	1.5	—	—
1988	50.4	17.1	10.5	6.2	0.4	3.6[b]	—	0.4	—	—	10.5
1991	61.4	17.7	1.8	2.1	1.1	—	—	0.6	—	8.3	4.4

Sources: Delal Baer, "The 1991 Mexican Midterm Elections," CSIS Latin American Election Study Series (October 1, 1991), 31; federal election data, Mexican Embassy, Washington, D.C.
[a] PRI = Institutional Revolutionary Party; PAN = National Action Party; PPS = Popular Socialist Party; PARM = Authentic Party of the Mexican Revolution; PDM = Democratic Mexican Party; PSUM = Mexico's United Socialist Party; PST = Socialist Workers Party; PRT = Revolutionary Workers Party; PMT = Mexican Workers Party; PRD = Democratic Revolutionary Party; PFCRN = Cardenista Front for National Reconstruction Party.
[b] Votes for the PSUM in 1988 were for the Mexican Socialist Party (PMS).

levels throve. Opposition party candidates were winning local-level executive positions and seats on municipal councils. PAN in particular was placing its members in important city posts in state capitals.

The declining legitimacy of the presidency, in combination with the declining legitimacy of the government itself, began to take a toll by the 1985 and 1986 elections. Although some of the smaller leftist parties were beneficiaries of these trends, the party capturing the greatest number of local seats was the PAN. The PRI worked hard to recoup the losses, in some cases by using such techniques as missing ballot boxes, duplication of registered voters, counting votes of citizens who had not voted, last-minute disqualifications of opposition poll watchers, and last-minute changes in the location of polling booths.[4]

The imposition of PRI victories on the subnational level reached a high point with the gubernatorial and local elections in the state of Chihuahua, a next-door neighbor of Texas. Ciudad Juárez, one of Mexico's largest cities, lies on the border and is traditionally a PAN stronghold. The PRI claimed victories in the state capital and in Ciudad Juárez, as well as in the state gubernatorial race.[5] The scope of fraud was so great and citizen resistance so palpable that prominent intellectuals and Catholic bishops denounced the results in a full-page ad in Mexico's leading daily, *Excélsior,* and called for the election to be annulled. As mentioned earlier, northern Churchmen announced they would cancel masses absent a government response. The Vatican delegate to Mexico, at the prompting of Mexico's secretary of government, persuaded the clergy to withdraw their threat.

The overall political environment laid the groundwork for the most significant change within the PRI, and led to the events that characterized the 1988 presidential election, a benchmark in Mexican electoral politics. Certain persons within the establishment leadership, in disagreement with the economic direction of the de la Madrid government and the timidness of his reforms, attempted a reform within the party's structure. In 1986 they constituted themselves as the Democratic Current. Among the most prominent members were Cuauhtémoc Cárdenas, a former governor of Michoacán and son of President Lázaro Cárdenas, the major political figure in the 1930s, and Porfirio Muñoz Ledo, a former cabinet official and president of the National Executive Committee of the PRI. The group and other party figures and intellectuals kept up a running criticism of the government's failure to implement genuine democratic reforms. Although at first tolerated, their dissenting voice became intolerable to the party and government leadership during the 1987 presidential succession. Party memberships were revoked.

Cárdenas, Muñoz Ledo, and other PRI dissidents formed the Democratic Front for National Reconstruction (FDN), selecting Cárdenas as its presidential candidate. Because the FDN joined the race too late to have its credentials legally recognized, one of the PRI's tiny splinter parties, the Authentic Party of the Mexican Revolution (PARM), selected Cárdenas as its nominee, thus giving the FDN a place on the ballot. The FDN's formation occurred at a time when internal contention over the selection of the PRI's candidate broke into the open. Carlos Salinas de Gortari, the programming and budget secretary, was seen by most observers as a man who would continue de la Madrid's economic philosophy; that suggested that the PRI's populist wing, represented by such individuals as Cuauhtémoc Cárdenas, would not have a presence in his administration.[6] The fact that Salinas had no prior electoral or grass-roots political experience marked the ascendancy of the technocratic leadership within the PRI.[7]

THE 1988 PRESIDENTIAL ELECTION

The 1988 presidential election illustrated a longtime pattern in electoral politics: the strongest opposition movements are often led by dissidents from within the PRI. As will be seen in the brief histories of several major opposition parties, most were founded by persons who abandoned government leadership because of policy and personal disagreements. This was true of the PAN and the Popular Socialist Party (PPS).

The 1988 presidential election occurred when the government and the PRI were at a low in terms of their legitimacy among the people. The selection of Salinas as the PRI candidate, the least popular choice among party leaders, further eroded the PRI's position. Given these conditions, the opposition began a vigorous campaign against Salinas. The PAN selected a charismatic businessman from the North, Manuel J. Clouthier, who provided energetic, if somewhat bombastic, leadership during the contest. Cárdenas was off to a rocky start, but with his name recognition, notably in rural Mexico, he began to build a following. Three leftist parties, which typically have attracted only small numbers of Mexican voters (see table 8-1), eventually joined Cárdenas's battle against the PRI candidate: the Popular Socialist Party (PPS); the Cardenista Front for National Reconstruction Party (PFCRN), formerly the Socialist Workers Party (PST); and the Mexican Socialist Party (PMS). Of the eight parties on the 1988 presidential ballot, four supported Cárdenas.

To most analysts' genuine surprise, the populist and leftist Cárdenas alliance generated a widespread response among Mexican voters. Cárdenas, according to official tallies, received 31 percent of the vote, the highest figure given to an opposition presidential candidate since the Revolution; Salinas obtained 51 percent, barely a simple majority; and Clouthier captured 17 percent, the typical PAN percentage in a presidential election. Contrary to most observers's expectations, the Left, not the Right, altered the face of the election. In other words, the 1982 voters who defected from the PRI six years later cast their ballots for Cárdenas, not the PAN.

It is important to remember that the 1988 election took place under the 1986 law, which allowed, for the first time, the majority party (the PRI) to increase its representation in the Chamber of Deputies from the 200 plurinominal seats. The PRI needed to implement the provision because it obtained only 233 majority district seats, 18 short of a simple majority of 251. It gave itself some 27 plurinominal seats, which added to the 233 majority district seats gave it a slight majority (260 out of 500). Table 8-2 illustrates the extraordinary shift in the parties' representation in the legislative branch. Up to 1988 the highest percentage of seats obtained by the opposition, combining majority districts and plurinominal seats, was 28 percent. In 1988, however, the opposition achieved 48 percent of the total, a 71 percent increase in three years.

Many observers of the 1988 election believe that the PRI engaged in fraudulent practices. Some—PRD figures among them—believe that Cárdenas actually won. Most, however, although agreeing with charges of fraud, also take into consideration the delays in reporting vote tallies and believe that Salinas actually did win—but that his percentage of the total vote was lower.

The 1988 elections appeared to suggest the end of Mexico's one-party-dominant system, the increased importance of pluralism in the political culture, and, as suggested in chapter 7, the greater importance of the legislative branch, where the PRI would have to negotiate with the opposition to obtain alliances sufficient to gain passage of legislation. Although the 1988 elections were a departure from a pattern, the 1991 congressional elections dampened expectations of an augmentation of opposition strength.

The PRI claimed 61.4 percent of the vote in the 1991 elections, a step back from 1988's brink but still within the pattern of decline since 1961 (see table 8-1). These elections wrought changes. First, whereas the PRI was on the verge of becoming merely a plurality party, controlling the electorate through the largest number of votes rather than exceeding a majority, it recovered its majority status. Second, the PAN basically remained at the same level as 1982 and 1988, retaining its position as the

Table 8-2 Seats in the Chamber of Deputies by Party, 1949–1991

Year	PRI	PAN	PPS	PARM	PDM	PSUM	PST	PRT	PMT	PRD	PFCRN
1949	142	4	1	—	—	—	—	—	—	—	—
1952[b]	152	5	2	—	—	—	—	—	—	—	—
1955	155	6	1	—	—	—	—	—	—	—	—
1958[b]	153	6	1	1	—	—	—	—	—	—	—
1961	172	5	1	—	—	—	—	—	—	—	—
1964	175	2	1	—	—	—	—	—	—	—	—
Party	—	18	9	5	—	—	—	—	—	—	—
1967	177	1	0	0	—	—	—	—	—	—	—
Party	—	19	10	5	—	—	—	—	—	—	—
1970	178	0	0	0	—	—	—	—	—	—	—
Party	—	20	10	5	—	—	—	—	—	—	—
1973	189	4	—	1	—	—	—	—	—	—	—
Party	—	21	10	6	—	—	—	—	—	—	—
1976	195	—	—	2	—	—	—	—	—	—	—
Party	—	20	12	9	—	—	—	—	—	—	—
1979	296	4	—	—	—	—	—	—	—	—	—
Plurinominal	—	39	11	12	10	18	10	—	—	—	—
1982	299	1	—	—	—	—	—	—	—	—	—
Plurinominal	—	50	10	0	12	17	11	—	—	—	—
1985	289	9	—	2	—	—	—	—	—	—	—
Plurinominal	—	32	11	9	12	12	12	6	6	—	—
1986[c]	233	38	4	5	—	—	—	—	—	15	5
Plurinominal	27	63	27	23	—	—	—	—	—	11	46
1991	290	10	—	—	—	—	—	—	—	—	—
Plurinominal	31	80	12	14	—	—	—	—	—	40	23

Source: Adapted from Héctor Zamitiz and Carlos Hernández, "La composición política de la Cámara de Diputados, 1949–1989," *Revista de Ciencias Políticas y Sociales* 36 (January–March 1990): 97–108.

[a] PRI = Institutional Revolutionary Party; PAN = National Action Party; PPS = Popular Socialist Party; PARM = Authentic Party of the Mexican Revolution; PDM = Democratic Mexican Party; PSUM = Mexico's United Socialist Party; PST = Socialist Workers Party; PRT = Revolutionary Workers Party; PMT = Mexican Workers Party; PRD = Democratic Revolutionary Party; PFCRN = Cardenista Front for National Reconstruction Party.
[b] Three other seats were won by members of the Federación de Partidos Populares Mexicanos and the Partido Nacionalista Mexicano.
[c] Three deputies were classified as independents, and one deputy among the PRD majority transferred his loyalty to the PRD after being elected on the PRI ticket.

second-ranking party, the key opposition party for three decades. Third, the PRD, the major beneficiary of PRI defection in 1988, was the primary source of PRI returnees in 1991. The PRD did not win a majority of votes in a single state in 1991, including those it had dominated in 1988 (Michoacán, México, Morelos, and the Federal District). Overall, its vote declined from nearly a third in 1988 to only 8 percent. Finally, instead of increasing the legislative branch's importance, the PRI's renewed dominance reinforced its weakness.

In the 1988 and especially the 1991 elections, Mexico's proximity to the United States played a special role. In 1988 the government became increasingly sensitive to charges of pervasive election fraud from abroad, especially from the United States media.[8] By 1991 Mexico's conduct in the electoral arena came under severe scrutiny in the U.S. Congress and in the media as discussions of the North American free trade agreement continued. American critics of the agreement charged that the United States should not be a party to it in light of Mexico's antidemocratic practices.

Despite the backtracking immediately prior to 1991, the pluralization of the political culture will go on. As Delal Baer argues,

> Two aspects of the 1980s democratic ferment are unlikely to be reversed. The first is the relatively greater importance of elections in the legitimation of power. The second [is] the more rigorous public expectations about the cleanliness of the electoral process. Electoral norms are pervasive from the medida to the local cafe. The mass protests of Guanajuato and San Luis Potosí are proof of a new civic culture that will continue to demand a clean electoral process.[9]

TRENDS IN MEXICAN ELECTIONS: OPPOSITION FORTUNES

As noted in chapter 7, many variables affect voters' perceptions of the parties. However, if we analyze the Mexican opposition in the aggregate rather than analyze each component party, we discover some universal characteristics. Among the variables most influencing opposition strength since the 1940s are location, relations with the government, level of development, and urbanization.[10]

Opposition to the PRI and its antecedents has a long history, despite the predominance of the government party since the late 1920s. Opposition has often coalesced around individual candidates and temporary party organizations, such as those of José Vasconcelos in 1929 and General Miguel Henríquez Guzmán in 1952, but opposition tendencies have remained strong in certain states and regions for many decades, indicating a permanence extending beyond any single issue or personality. For example, examination of the 1946, 1952, 1982, and 1988 presidential elections, when the opposition was strongest, indicates that the opposition consistently obtained at least 30 percent or more of the vote in seven states: Baja California, the Federal District, Guanajuato, Jalisco, México, Michoacán, and Morelos. In the 1991 elections, except for Jalisco and Morelos, the opposition attracted more than 40 percent of the vote; and in the Federal District and Baja California, more votes than the PRI.

Table 8-3 PRI Vote by State Per Capita Income Level, 1976–1991

Jurisdiction[a]	Election Year					
	1976 (%)	1979 (%)	1982 (%)	1985 (%)	1988 (%)	1991 (%)
Nation	87	74	68	68	50	61
Low-income states	92	81	82	81	75	71
Medium-income states	93	78	73	68	56	66
High-income states	78	55	55	52	37	53

Sources: Adapted from Paulina Fernández Christlieb and Octavio Rodríguez Araujo, *Elecciones y partidos en México* (Mexico: El Caballito, 1986), 218, 223–24; Joseph Klesner, "Changing Patterns of Electoral Participation," in *Mexican Politics in Transition,* ed. Judith Gentleman (Boulder, Colo.: Westview Press, 1987), 130, 135; *El Día,* July 16, 1988, 8; Instituto Federal Electoral, 1991 election data, district computations, courtesy of Luis Medina, Mexican Embassy, Washington, D.C.

[a]High-income states-Baja California, Baja California Sur, Chihuahua, Federal District, México, Nuevo León, and Sonora; low-income states = Chiapas, Hidalgo, Oaxaca, Puebla, San Luis Potosí, Tlaxcala, Yucatán, and Zacatecas; medium-income states = the remaining states.

Many reasons exist for the long opposition to the PRI in the named states. For some—Baja California, the Federal District, and México—growth, income, urbanization, and development are important; they are among the six states with the highest per capita income. (see table 8-3) Economic growth is likely to have the most influence on opposition strength over the long term. It has been argued that economic growth and development are statistically related to election results.[11] In the United States we know, for example, that when the economy does poorly, support for the president and his party generally declines; this was so for President Bush in the fall of 1991 and into 1992. In Mexico, however, although strong economic performance in the short term increases voter satisfaction with the government, when economic performance translates into higher standards of living, those who live in regions benefiting most vote for the opposition, contrary to expectations.

The region of the country that has best demonstrated the surprising voting behavior over time is the Federal District, which has always been among the high-per-capita-income states, and has always recorded sizeable votes for the opposition. The PAN has done extremely well in the capital for many years, as did the PRD during the 1988 presidential elections. To illustrate the relationship between economic development and support for the opposition, we can group Mexico's states according to per capita income, and the percentage of votes for and against the PRI (see table 8-3). In each election year since 1976, whether we are dealing with presidential or legislative candidates, the total percentage of votes for the PRI in high-income states is much lower than the national average. On the other hand, the PRI obtains a disproportionate amount of its support from low-income states.[12]

Why is the opposition stronger in states benefiting economically under PRI leadership? A number of reasons stand out. One, the PRI is much better organized in rural areas, and low-income states typically are the most agrarian. Two, educated Mexicans, who are more sophisticated about participation and more likely to vote for an opposition candidate, live in greater numbers in high-income states. Three, supervision of voting in urban centers, often located in the high-income states, is characterized by fewer reports of fraud and hence fairer.

Long-term development trends, therefore, tend to favor the opposition, not the government's party. This pattern has implications for policy because the federal government, which collects more than 85 percent of tax revenues, decides how they are to be distributed. In terms of direct investment, the government typically has favored the most economically developed states. Ironically, then, the government has reinforced the pattern by directing financial goods to the regions that support the PRI the least.[13]

Geographic location is important to other states. For Baja California, proximity to the United States has played an important role. It is difficult to link empirically that proximity with Mexican voters' support for the opposition. There is little doubt, however, as we suggested earlier, that where Mexicans live affect their views of the United States and its political system, and their political values.[14] In fact, surveys show that the closer Mexicans live to the United States and the more they travel to it, the greater their admiration.[15] Many of the prominent figures in the PAN are products of the border culture, including its 1988 presidential candidate, who attended high school in San Diego. In fact, Mexicans living in the United States repeatedly have requested the right to vote by absentee ballot, a request consistently denied. Observers argue that the government refusal is based on the high probability that the overwhelming majority would support opposition parties.[16]

Historical experience has also had much to do with the level of opposition support in various states, generally in connection with poor relationships. For example, Morelos, a state just south of Mexico City, gave birth to the most important agrarian movement during the 1910 Revolution, led by Emiliano Zapata. Zapata and his followers never pursued goals compatible with those of the Mexican government, even after the initial success of the Revolution in 1911, or its more institutional phase after 1916. Hence, tensions have always existed between the peasants in Morelos and national political leadership.

Strong opposition in Guanajuato, Jalisco, and Michoacán stem from the importance of Catholicism, and Church-state relations in those states during the 1920s. When the federal government strictly applied constitutional provisions relative to religion, it provoked resistance from staunch Catholics and some clergy in the region including the three states. This movement in defense of religious rights became known as the Cristero rebellion. Memories of the events are still fresh in the minds of the people who were children in the 1920s. The region contributes heavily to the priesthood and hierarchy, and a disproportionate number of bishops are graduates of the Morelia seminary in the heart of Michoacán.

Geography, level of development, income, and urbanization, among other variables, functioned to add to the historic importance of opposition. Although none of the opposition parties yet has strength enough to best the PRI nationally, the foundation to do so has existed for many decades. The time will come when one of them, or some new opposition party, will effectively contest for primacy among the electorate.

OPPOSITION PARTIES: THEIR ORIGINS AND FUTURE

The most important opposition party in Mexico, and the longest-lived in terms of putting up promising candidates is the National Action Party, founded in 1939 by Manuel Gómez Morín, a former national figure in government economic policy-making, and Efraín González Luna, a lawyer and Catholic activist. As is true of so many of the opposition parties, leadership often came, at least initially, from disgruntled establishment elites. In some cases it is individuals who have had their own political ambitions cut short; in other cases it is a question of policy differences. The PAN's formation is an example of the latter's bringing together diverse individuals who were against the statist, populist economic policies of President Lázaro Cárdenas (1934–1940).

The PAN first put up candidates against the PRI on the local level. Although it had supported several opposition presidential candidates, it did not run its own candidate until 1958. It has put up a candidate in every election since then except 1976, when it protested the PRI monopoly and electoral fraud by withdrawing from the presidential contest. Ten years after its founding, it captured less than 3 percent of the legislative seats (table 8-2). After the first electoral reforms went into effect in 1964, it began obtaining roughly 10 percent of the seats, assisted by its share of party deputies. Its representation stabilized around that figure until 1988, when it doubled its share to 20 percent. In the 1991–1993 legislature it fell to 18 percent.

The PAN ideological banner has shifted over time. Initially, the party leader, many of whom were well connected financially or had ties to Catholic Action youth or other Catholic movements, were described as conservative, in some cases reactionary, probusiness, and pro-Church. The party evolved gradually into a Mexican variant of a Christian Democratic organization by the 1960s.[17] However, like the PRI, and especially the left-of-center parties, the PAN suffered from internal dissent. At various points its leadership has wavered between those desirous of playing more or less by the rules of the political game, as set down by the government, and those who advocated a more aggressive stance vis-à-vis the PRI.

The new PAN activists, sometimes referred to as neo-Panistas, have taken a more conservative stance ideologically, and have often allied themselves with combative businessmen willing to run under the PAN banner. As Soledad Loaeza suggested in her analysis of the 1988 presidential election, the neo-Panistas broke new ground "by challenging the unwritten rules of Mexican politics—in particular, the idea that industrialists should not participate in politics."[18] When President Salinas took some of the thunder out of traditional PAN issues like statism, labor corruption, and outmoded Church-state relations, the PAN centered its major criticism on political modernization, notably genuine electoral reform, a popular issue with the electorate. Yet PAN's leadership, even while faced with vehement internal dissension, eventually supported an electoral law introduced by Salinas, a law it believed would ensure integrity in voter registration and balloting. It also introduced the so-called governability clause, which guaranteed 51 percent of the seats in the Chamber of Deputies to the majority party obtaining at least 35 percent of the vote. The 1989 reform also made it extremely difficult for smaller parties to put up coalition candidates, as Cárdenas had done under the National Democratic Front in 1988.

The PAN's growth nationally has never been dramatic. It has remained the major opposition party, except during the 1988 presidential race, but its strength since 1982 has stabilized at approximately 17 percent. National figures and figures from each of the three hundred majority districts suggest the regional quality of PAN support. Its organizational strength and the narrowness of its platform have made it viable primarily in urban centers. Indeed, it would be accurate to call the PAN a regional, urban party. Even in 1988, the only time that the PAN won more than 4 percent of the legislative majority districts, the 38 seats it obtained were in major cities, including Ciudad Juárez; Mexico City; León; Guadalajara, the capital of Jalisco; several districts in the industrialized section of the state of México; San Luis Potosí and Mérida, both state capitals; and Culiacán. In fact, four-fifths of its legislative seats were in just five Mexican cities.

The PAN's potential for the future, even with more stringently applied electoral reform, is stronger at the local and state levels than nationally. One region where this growth is likely to occur is the North, where many owners of small and medium-size businesses are pressing for a more open voice in partisan politics.[19] As data in the previous chapter indicate, only a small percentage of prospective voters express sympathy for the PAN. A stronger national future is also made less likely by the fact that at least under Salinas, PRI's ideology has absorbed many of its economic ideals, rendering PAN's program less distinctive. People who vote for the PAN do so because they want a change from the PRI.[20] Nevertheless, the PAN obtained the governorship of Baja California in 1989, the first such victory officially recognized by the government. It reinforced its appeal in that state by obtaining the Senate seat in the 1991 elections. PAN complaints about alleged fraud in the Guanajuato gubernatorial race in 1991, echoed in the U.S. media, led the government to force the resignation of the PRI candidate. He was replaced by an interim governor, also a Panista. If the PAN can win the executive branch in more states, as it has done on the local level since the mid-1980s, its grass-roots organizations would prosper, which could lead to a stronger performance in more majority districts for congressional seats. It strengthened its potential in 1992 by adopting a new strategy locally. In Durango, it joined the PRD in an electoral alliance. In exchange for supporting PAN's gubernatorial candidate, PAN supported PRD's candidate for mayor of the state capital. In the immediate future, however, the PAN is not likely to garner more than 20 to 25 percent of the national vote.

The only other major opposition is the Democratic Revolutionary Party. The PRD has a short history; it constructed itself on a foundation of smaller leftist parties that had flowered during the 1970s. Elements of the Mexican Communist Party (PCM), founded in 1919, and the Mexican Socialist Party (PMS), founded in 1987, provided the formal organizational basis. The PRD came into being after the 1988 presidential election. Many of its founders, as in the case of the PAN, were PRI dissidents.[21] Some of its members in the Chamber of Deputies and in the Senate previously held political posts as Priistas, or as leftist-party members.

The PRD's ideology is difficult to characterize because the party's ranks are an amalgam of political groups professing views ranging from Marxist to populist. Some issues prominent in the PRD platform and that distinguish it from the PRI include the electoral reform of 1989, debt-payment renegotiation with U.S. and other foreign bankers, privatization, land distribution, exchange rates, and political reforms generally. In short, the PRD advocated more radical negotiations favorable to Mexico on the debt;[22] preached caution on privatization; opposed the 1989 electoral reform; favored land redistribution; advocated fixed exchange rates to protect working-class incomes; and called for thoroughgoing electoral reforms. Typically, it has favored economic policies that cater less to business interests, and advocated the traditional importance of the state in economic affairs.[23]

The results of the 1988 election were deceptive in terms of PRD strength. The immense popularity of candidate Cárdenas's father among many sectors of the population made it difficult to distinguish support for Cuauhtémoc Cárdenas as a symbol of his father from support for the principles of his coalition. Those who vote for the PRD, as in the case of the PAN, are primarily interested in change. But the PRD, even more than the PAN, given its origins in a loosely connected alliance of small parties, has found it difficult to maintain its cohesion. One of the parties in the original alliance, the Cardenista Front for National Reconstruction, has pursued an independent course since 1988.

Some observers have noted a decided strategy on the part of the PRI in its electoral contests with the PRD to maintain PAN strength and to focus on defeating the PRD. The strategy, it is argued, became apparent in 1989, after the PAN won the governorship of Baja California, and the PRI claimed surprising victories in local and state legislative districts in Michoacán, Cárdenas's home state. By 1991 PRD electoral strength had declined precipitously to only 8 percent of the vote, a distant third to the PRI and the PAN (table 8-1). The PRD faced electoral conditions involving fraud, and many of its active supporters were physically threatened, injured, or killed. In fact, PRD members constitute the single largest group of victims in national and international Human Rights Commission reports.

Voter support for the Left is even weaker nationally than voter support for the Right, expressed through the PAN. Voter strength in 1988, the high point of leftist opposition, suggests the diminution. At the legislative level the Cardenistas did not win nearly the proportion of seats its national vote for the presidency indicated as its potential strength. For example, it outpolled the PRI 49 percent to 27 percent in the Federal District but carried only 8 percent of the district's legislative seats. The same was true in the state of México, where the Cardenistas also were strong at the presidential level. The PRD's strength at both levels extended only to the state of Michoacán, and was not repeated in the 1991 elections. As Barry Carr had predicted, unless the leftist parties were able to translate their 1988 election victories into grass-roots labor and agrarian organizations, they would not create a base sufficient to sustain future electoral successes.[24] This is essentially what has happened. Because ideological appeals from the Left are not popular among the voters according to recent polls, the PRD, unless it shifts and redefines its program, is likely to have little appeal during the remainder of the decade.

CONCLUSION

Election trends in Mexico, with the exception of the 1988 presidential election, demonstrate some remarkably consistent patterns. Despite the continued monopoly of the PRI as the government electoral vehicle and the PRI resurgence in the 1991 elections, through both grass-roots efforts and alleged fraud, it has continued to decline slowly in overall support, leaving an opening for present and future opposition parties.

Economic modernization, if it impacts favorably on the standard of living of individual Mexicans as well as geographic regions, is likely in the long run to give impetus to support for opposition parties, given increased voter sophistication and the difficulties in perpetrating fraud in urban centers. Historically, higher-income regions, those benefiting most from government policies, have voted against the PRI, not for it. Although the PRI continues to appeal to a wide range of income groups, and is the only party demonstrating depth nationally in terms of its base of support, it has many vulnerabilities.

Two conditions prevent opposition parties from capitalizing on PRI weaknesses. The first is the relationship between the party and the government. Although government performance is generally associated in competitive, democratic systems with the success or failure of its party, Mexico's PRI can ride out these ups and downs economically and otherwise because of advantages gained through its symbiotic relationship to the government. The most important advantages are financial benefits, more indirect than direct, and the human resources that accrue to the party through the government. For example, in a presidential campaign, army troops and officers are assigned to the PRI candidate, and carry out many logistical activities. Furthermore, the PRI gains from tremendous coverage in the media; studies show that it receives far more attention from the press than do any of the other parties.[25] Its ability to furnish transportation to reporters, provide them with favors, and give them first-rate communication facilities all lead to PRI dominance over the media. Most important, PRI monopolization of government offices provides it with a reward system unmatched and unattainable by any other party. If the same kinds of "goodies" inhered in legislative positions, opposition parties could offer more attractive rewards to their leaders and rank and file. Instead, the rewards they can offer are few and less worthwhile politically.

The second inhibiting condition is intertwined with the first: the extraordinary difficulty attached to bringing a major opposition entity into being. From an ideological point of view, Mexico needs another centrist party, perhaps even center-right, that would accord with the desires of the electorate as revealed to pollsters. Even should such an entity materialize it would not be likely to sustain itself beyond a single election, given the first condition described above. Until a potential leader can capture the imagination of many voters, can appeal to the ideological center, and can appeal to all social backgrounds, the PRI political hegemony is secure. Only when elections are almost universally deemed honest and fair will citizens become more interested in the process and its outcome.

NOTES

1. For background, see John Bailey, "Can the PRI be Reformed," in *Mexican Politics in Transition,* ed. Judith Gentleman (Boulder, Colo.: Westview Press, 1987), especially, 74–94.

2. Wayne Cornelius, "Political Liberalization in an Authoritarian Regime: Mexico, 1976–1985," in *Mexican Politics in Transition,* ed. Judith Gentleman (Boulder, Colo.: Westview Press, 1987), 22.

3. María Emilia Farias Mackey, *The Reform of Electoral Policy* (Mexico: Congreso de la Unión, 1987), 8–9.

4. Daniel Levy and Gabriel Székely, *Mexico: Paradoxes of Stability and Change,* 2d ed. (Boulder, Colo.: Westview Press, 1987), 68.

5. For detailed background, see Delal Baer, "The 1986 Mexican Elections, The Case of Chihuahua," CSIS Latin American Election Study Series (Washington, D.C.: Georgetown University, September 1986).

6. For background, see Peter H. Smith, "The 1988 Presidential Succession in Historical Perspective," in *Mexico's Alternative Political Futures,* ed. Wayne Cornelius, Judith Gentleman, and Peter H. Smith (La Jolla: United States-Mexico Studies Center, UCSD, 1989), 399ff.

7. Roderic Ai Camp, "Mexico," in *Latin America and Caribbean Contemporary Record,* ed. James M. Malloy and Eduardo A. Gamarra (New York: Holmes & Meier, 1990), B300.

8. Leonardo Ffrench Iduarte, "The Mexican Presidential Election in the Mass Media of the United States," in *Sucesión Presidencial: The 1988 Mexican Presidential Election,* ed. Edgar W. Butler and Jorge A. Bustamante (Boulder, Colo.: Westview Press, 1991), 218.

9. Delal Baer, "The 1991 Mexican Midterm Elections," CSIS Latin American Election Study Series (Washington, D.C.: Georgetown University, October 1, 1991), 29.

10. See my "Mexico's 1988 Elections: A Turning Point for Its Political Development and Foreign Relations," in *Sucesión Presidencial: The 1988 Mexican Presidential Elections,* ed. Edgar W. Butler and Jorge A. Bustamante (Boulder, Colo.: Westview Press, 1991), 98ff.

11. Barry Ames, "Bases of Support for Mexico's Dominant Party," *American Political Science Review* 64 (March 1970): 153–67.

12. For the most comprehensive analysis of these and other variables, see Joseph Klesner, "Electoral Reform in an Authoritarian Regime: The Case of Mexico" (Ph.D. diss., MIT, February 1988); for 1988, Antonia Martínez Rodríguez, "Que hable México: último gobierno priista?" *Revista de Estudios Políticos* 63 (January–March 1989): 251–58.

13. Roderic Ai Camp, "A Reexamination of Political Leadership and Allocation of Federal Revenues in Mexico, 1934–1973," *Journal of Developing Areas* 10 (1976): 100–102.

14. Alberto Hernández Hernández, "Political Attitudes among Border Youth," in *Electoral Patterns and Perspectives in Mexico,* ed. Arturo Alvarado (La Jolla: Center for U.S.-Mexican Studies, UCSD, 1987), 216.

15. William J. Millard, *Media Use by the Better-Educated in Major Mexican Cities* (Washington, D.C.: U.S. International Communications Agency, 1981); Alberto Hernández Medina and Luis Narro Rodríguez, eds., *Cómo somos los mexicanos* (México: CREA, 1987).

16. Comments, Symposium on Mexican Electoral Reform, Institute of Latin American Studies, University of Texas, March 1991, Austin.

17. Donald Mabry, *Mexico's Acción Nacional: A Catholic Alternative to Revolution* (Syracuse: Syracuse University Press, 1973).

18. Soledad Loaeza, "The Emergence and Legitimization of the Modern Right, 1970–1988," in *Mexico's Alternative Political Futures,* ed. Wayne Cornelius, Judith Gentleman, and Peter H. Smith (La Jolla: Center for U.S.-Mexican Studies, 1989), 361.

19. See Yemile Mizrahi, "The New Conservative Opposition in Mexico: The Political Radicalization of Northern Entrepreneurs" (Paper presented at the National Latin American Studies Association meeting, April 1991).

20. Gallup Organization, exit poll, August 18, 1991. This is the first exit poll permitted by the Mexican government.

21. The evolution of the Democratic Current and the formation of the original electoral front, as described in detail by Cuauhtémoc Cárdenas in a lengthy interview with Carlos B. Gil, *Hope and Frustration: Interviews with Leaders of Mexico's Political Opposition* (Wilmington, Del.: Scholarly Resources, 1992), 155ff. This is the first work of its kind on Mexico, and provides a broad sense of opposition views.

22. For clarification on this specific issue, see Jesús Galindo López, "A

Conversation with Cuauhtémoc Cárdenas," *Journal of International Affairs* 43 (Winter 1990): 395–406.

23. Javier Farrera Araujo and Diego Prieto Hernández, "Partido de la Revolución Democrática: documentos básicos," *Revista Mexicana de Ciencias Políticas y Sociales* 36 (January–March, 1990), 67–95. For a comparison with PAN and other parties, see Federico Reyes Heroles, ed., *Los partidos politicos mexicanos en 1991* (Mexico: Fondo de Cultura Económica, 1991).

24. Barry Carr, "The left and Its Potential Role in Political Change," in *Mexico's Alternative Futures,* ed. Wayne Cornelius, Judith Gentleman, and Peter H. Smith (La Jolla: Center for U.S.-Mexican Studies, 1989), 381.

25. See "Espacio a los partidos en la prensa, 1988–1990," *El Nacional,* June 22, 1991, 1.

9

Political Modernization:
A Revolution?

> [E]ven though democratic rules are necessary conditions for political alternation, they do not necessarily result in an alternation in power, unless we resort to non-democratic "corrective" actions.
>
> Nevertheless, as long as the government continues to be seen as the only agent capable of managing and making decisions on electoral results, and as such, as the only interlocutor of contending political parties, a view inherited from years of centralized decisions, the electoral process will remain a source of public skepticism.
>
> ROBERTA LAJOUS, *Examen*, November 1991

The political question foremost in the minds of most observers and Mexican citizens alike is, What will be the influence of liberalism, economic and political, on Mexico's governmental model and its economic future? Many of the political and economic changes that have occurred in eastern Europe since 1989 have far exceeded the expectations of most experts. Boundaries have been redrawn; political processes have been turned upside down; regional stability has been rendered problematical; and economic structures have collapsed or are tottering. Mexico has not been immune to the winds of change. Some analysts argue that international influences and world public opinion have strongly affected Mexico. Mexican architects of recent reforms, including Treasury Secretary Pedro Aspe, consider Mexico to be part of "this vast process of world institutional evolution."[1]

ECONOMIC LIBERALIZATION

Whether the source of economic change in Mexico is international or domestic or both, there is no question that the administration of Carlos

Salinas de Gortari has instituted major economic reforms indicative of an altered government economic philosophy. As we argued earlier, the relationship between the state and the private sector has long been symbiotic and contentious. Its features emerged from a hybrid economic philosophy extending back to nineteenth-century laissez-faire liberalism and to social responsibility themes incorporated into the 1917 Constitution.

For most of the twentieth century, the government offered a mixed private-public economic model, wherein the state played a decisive, sometimes overpowering role. In the 1970s, under President Luis Echeverría (1970–1976), the legitimacy of the political model increasingly came into question. Echeverría's difficulties, in part, stemmed from the events of 1968, during which the army violently suppressed a student demonstration in the capital. He pursued various social and economic strategies to enhance presidential legitimacy and the political model's prestige.[2] Public employment was used to foster economic growth and stability, increasing opportunities for many Mexicans. One social scientist remarked, "Even though the public sector had had a relatively dynamic growth throughout the previous decade, during the first half of the 1970s it broke all precedents."[3]

During Echeverría's tenure large reserves of petroleum were discovered. The government embarked upon exploitation of those reserves and on the basis of oils sales abroad secured international loans to finance development projects.[4] During the early 1970s the government bought or gained control of hundreds of businesses and industries, placing more economic and human resources in the hands of government managers than at any time theretofore. At the end of his administration Echeverría further alienated the private sector by attempting to expropriate valuable lands in the Northwest.

Echeverría's successor, José López Portillo, a politician-technocrat experienced in the government financial sector, attempted to mend relations between the private and public sectors. Initially successful, he continued the pattern of borrowing large sums of money to invest in Mexico's economic infrastructure and development. When the oil boom abruptly ended and prices at the barrelhead plunged, the country found itself in serious trouble. It was hugely indebted to domestic and foreign bankers, and paying extraordinarily high interest.[5]

López Portillo, instead of putting the brakes on the state's economic expansion actually stepped on the accelerator. In his last year in office, without warning or consultation, he announced the nationalization of the domestic banking system.[6] With a single decree, the president increased state control over the economy, of which bank-held mortgages increased it to somewhere between 75 and 85 percent. The move exacerbated the business community's lack of trust in the government and strongly encouraged the flight of capital from Mexico, primarily to the United States.[7] When López Portillo left office a few months later, the presidency was at its lowest ebb in decades; the business-government relationship was in great disrepair; and Mexico was in economic crisis.

López Portillo's successor, Miguel de la Madrid (1982–1988), like his mentor, was a product of the public financial sector, having worked in Mexico's equivalent of the U.S. Federal Reserve Bank, in the Secretariat of the Treasury, and as secretary of planning and budgeting. His economic philosophy, however, represented a different ideological wing of the government leadership. Essentially, he believed that the most likely strategy for rescuing Mexico from economic woe was to follow the strict, orthodox economic guidelines recommended by the International Monetary Fund (IMF): to reduce government expenditures, and impose controls on salaries, prices, and inflation.[8]

De la Madrid also introduced the most important element in Mexican economic liberalization: privatization. He actually wanted to undo the nationalization of the banks—which in itself would have had an immediate impact on government ownership—but believed that the mid-1980s were not a politically propitious time for the move.[9] Instead, he initiated several moderate steps that made possible joint private-public ownership of certain financial institutions. At the end of his administration, it became clear that some government-owned firms would be sold back to the private sector.

De la Madrid ensured the importance of privatization specifically and economic liberalism generally in selecting Salinas to follow him. The 1988 presidential succession took on great significance within the political leadership. In one sense, competition among contenders within the PRI represented a conflict between a more traditional economic philosophy, which favored state control and deficit-spending budget strategies, and the more orthodox private-sector emphasis that de la Madrid had reintroduced.[10] Although de la Madrid had improved the relationship between the private sector and the state, Salinas by his second year in office, had established a clear-cut policy incorporating many ingredients of international economic liberalism.

It is very important to take note of U.S. influence on the Salinas policy. The United States does not play a direct role in the formulation of Mexico's economic policy. Nevertheless, Reagan pushed and now Bush pushes a more orthodox economic policy domestically and similar policies elsewhere, including Mexico. Throughout the 1980s the United States expressed serious concern about Mexico's stability, and its economic and political future. The American financial community, which held large portions of the Mexican government's debt portfolio, echoed the concern. Default might well have initiated a Latin American domino effect, with drastic consequences for the already shaky U.S. financial structure and the U.S. economy.[11] Salinas saw capital as essential to Mexican economic recovery in the short term and international competition in the long term. When he realized that European governments and lenders were preoccupied with eastern Europe, he turned to a free trade agreement with the United States and Canada. Bush, who has close ties to Salinas, committed himself and the United States to approval of such an agreement and encouraged Salinas to move ahead on it. In anticipation, Salinas and his economic team, most of whose members studied in the United States, began to put many government-owned firms up for sale and to cut tariffs dramatically—many dropped from as high as 200 percent to an average of only 9 percent in 1992. The initiatives led to the return of some domestic capital and to new foreign investment—more than $10 billion by 1991.[12]

In late 1991 and early 1992 the government began to sell off the banks it had nationalized a decade earlier. It also put several major corporations owned by the government on the market, including Teléfonos de México (Telmex), which has a monopoly on telephone communications in Mexico, and Mexicana Airlines, one of the two major domestic lines. In fact, of the 1,155 firms that the government owned as late as 1987, it retained control of only 286 in 1992, a drop of 80 percent.[13]

The U.S. financial community responded favorably to these dramatic changes from a state-led to a free-market economy. Editorials in business-oriented publications like the *Wall Street Journal* praised Salinas and his collaborators. Other periodicals, such as *Business Week,* predicted a boom period for Mexico, making it attractive to investors. The Mexican government repeatedly cited its positive press as evidence supportive of its policies.

Yet critics charged that the state-controlled sector remained bloated. They argued that twelve of the twenty largest firms in terms of employees were still under state control. Indeed, that state-owned firms employed 79 percent of all workers, and that the size of the bureaucracy was not much reduced from what it had been in 1987—from 4.4 to 4.1 million, largely the result of the sale of banks and government-owned companies.[14]

Parallel to his commitment to privatize and open up the economy to international competition, Salinas in 1992 startlingly proposed to overhaul the *ejido* land structure, a system of small-property holding controlled by each village. After the Revolution successive governments gave lands to individually operated ejidos, whose owners received use-right titles but not actual ownership from their local villages. Critics of agrarian reform charge that insufficient credit stems from an ejidatorios's lack of collateral. Salinas's new legislation grants actual ownership and contract rights to these farmers.

Salinas introduced one other major social-economic policy soon after taking office, one closely linked to political liberalization. Known as the National Solidarity Program (Pronasol), or popularly as "Solidarity," it provided government seed money for local projects. Ostensibly, the philosophy behind the program was to encourage grass-roots organization and local leadership. Thousands of farmers received loans, communities established rural medical clinics and renovated schools, and scholarships were awarded to promising students. Its supporters assert that it promoted grass-roots organization and leadership because local residents chose and prioritized the programs. Critics in opposition parties and some independent observers view Pronasol as a sophisticated, centrally controlled funding agency that has built considerable electoral support for the government party since 1989. Still others believe that it is a means of enhancing the personal power and political influence of the president.[15] The president institutionalized this new program, giving it cabinet status in the new Social Development Secretariat, appointing the former head of the PRI, Luis Donaldo Colosio, to direct it.

Regardless of the weaknesses and strengths of the economic policies of Salinas, he has pursued a consistent economic strategy, composed primarily of privatization, internationalization, and foreign investment. As part of his overall strategy to modernize Mexico, he is personally much more committed to economic than political liberalization. In fact, as he has made clear, he plans to pursue modernization and, implicitly, the interrelationship of the two components. Not wishing to make the same errors as had the former Soviet Union, Salinas declared:

> Freedoms of what you call the glasnost kind have existed for decades in Mexico. What hasn't existed is the freedom of productive activity, because the government owned so many enterprises.
>
> So, actually, we have been more rapidly transforming the economic structure while striking along many paths of reform on the political side.
>
> But, let me tell you something. When you are introducing strong economic reform, you must make sure that you build the political consensus around it. If you are at the same time introducing additional drastic political reform, you may end up with no reform at all. And we want to have reform, not a disintegrated country.[16]

Salinas carefully planned his economic strategy and, more important, gave particular attention to the constituencies to whom it was addressed. His prudence paid off in considerable domestic and foreign support. He also used his economic successes—higher capital investment, lower inflation rates, and increased dollar reserves—to mollify some of his domestic and many of his international critics, especially the United States.[17]

It also can be argued that Salinas used his economic successes to bolster his political prestige. Unlike the goals of political modernization, one of which is the decentralization of authority, the results of the successes, according to some critics, were a strengthened presidency, enhanced centralized decision making, reduced electoral competition, and a leaner, stronger state.[18]

DEMOCRATIZATION

In Mexico, as elsewhere in the world, political liberalization has meant democratization. Thus, the international winds of change indicated a combined political-economic model, incorporating political democracy on one hand and economic capitalism on the other. Mexicans had long expressed an interest in democratization, which flared up after independence, in the 1860s and 1870s, at the time of the Revolution, and then again in the 1960s and 1970s.

As was true with the movement toward economic liberalization, President de la Madrid paved the way toward recent political events. Soon after taking office, as part of a moral renovation, he promised cleaner elections and decentralization of the candidate selection process within the government party. Initially, the promises translated into actual improvements, and opposition parties, specifically the National Action Party, won many local elections in the mid-1980s.[19] Surprised, government officials reversed the apparent opening by resorting to the "doctoring" of vote counts. The practice roused the Catholic hierarchy and the country's leading intellectuals to proclaim disbelief in government vote counts in the 1986 elections in Chihuahua.

Within the government leadership, a debate ensued as to future political and economic strategies. The debate centered on two primary, interrelated issues. First, should the government expand its commitment to the economic programs introduced gradually under de la Madrid, programs that reduced the standard of living of a fourth of Mexico's economically active population, or should it resort to spending programs to moderate the drastic effects of austerity, and resist paying the international debt? Second, should the leadership open up the political system to widespread competition and report electoral results honestly or continue along the same road? The political figures who wanted to return to deficit spending, economic nationalism, and strong state leadership, combined with electoral honesty, lost out in the internal battles for the presidency. When they tried, particularly on the issue of democratizing the PRI, to pressure the leadership from within, they were summarily dismissed from PRI party ranks in 1987. Their decision to form an opposition movement under Cuauhtémoc Cárdenas provided a catalyst for the most important election in recent Mexican political history.

The presidential election of 1988 was a test of the legitimacy of the establishment leadership. Although preelection polls indicated the strength of Cuauhtémoc Cárdenas, most analysts underestimated his appeal to the electorate. It was clear that Salinas was the least popular choice among his own party rank and file, and he personally generated little additional support during the campaign. The official results show that Salinas won 50.74 percent of the vote; the Cárdenas's alliance, 31.06 percent; and Manuel Clouthier, the National Action Party's candidate, 16.81 percent.

The official results were widely disputed in the media. As noted in the previous chapter, many independent observers and critics believe the figures were fraudulent, and many suggested that Cárdenas may actually have defeated Salinas. Most Mexicanists contend, and they are borne out by survey research, that Salinas won but with a much smaller margin than actually reported.[20] Even accepting the official results, the PRI and the establishment leadership gave up more seats in the Chamber of Deputies and, for the first time, in the Senate than at any other point theretofore.

The election of 1988 had numerous consequences. Among them, it provided a catalyst for the development of a new opposition party, the Democratic Revolutionary Party (PRD), which slipped from second to third place in 1991 but remains an important alternative to the PRI and PAN. The PRD was formed in 1988 after the demise of a temporary electoral alliance under Cárdenas. Second, the election gave greater prominence to the Chamber of Deputies, where nearly half of the seats, proportional and district alike, went to the opposition. Third, it brought the democratic desires of the populace to the fore and gave much greater visibility to desires that had been expressed electorally on the district and regional levels since the 1940s but never so dramatically in a national election.[21] Fourth, it contributed to new political alliances, notably between the PAN and the PRD in the electoral arena, and between the PAN and the PRI in the policy arena. Fifth, it paved the way for a series of gubernatorial elections, directly or indirectly leading to opposition victories in 1989, 1991, and 1992. Finally, it forced the government leadership to reformulate its political constituencies, and in doing so, introduced political changes as dramatic as those in the economic sphere.

Salinas took office in December 1988 with only a minimal level of political legitimacy. Having won or imposed the disputed election results, he faced—as no president in recent memory had faced—inauguration with very little public support. Salinas confounded his detractors and supporters alike, moving quickly to establish a reputation as decisive. Instead of depending on the presidency to increase his political influence, Salinas enhanced the prestige of the presidency with his own power and strength. He did this in a series of deft decisions, including using the army to arrest a corrupt union leader and to seize a major drug trafficker, as well as arresting and prosecuting a well-known businessman for financial fraud.

Throughout his tenure Salinas used the presidency as the leading institution to implement his policies and to centralize control of decision making. One of the more interesting ironies of his administration is the concentration of decision-making authority in the presidency.[22] He has streamlined the workings of the cabinet removing the agency that both he and his mentor, de la Madrid, had used to rise to the top of the political ladder. He eliminated programming and budgeting by joining it with the Secretariat of the Treasury. He further coordinated cabinet planning by expanding interagency planning through subcabinet groups.

At first glance, Salinas's promise of freer elections appeared to be translated into victories for the opposition. In Baja California, where opposition parties had dominated in the 1988 presidential election, the PAN won its first gubernatorial contest in 1989. No opposition candidate had been permitted to win since the 1930s. It became clear, however, that the government was not committed to political modernization to the same degree as economic liberalization.

In 1991 two important elections took place in San Luis Potosí and Guanajuato. Both states have long histories of support for opposition parties, notably the PAN, but San Luis Potosí generated its own opposition movement in the 1950s.[23] In both cases, the elections were contentious, but the PRI claimed overwhelming victory in each. What most observers found difficult to credit was the *level* of the PRI victory. In San Luis Potosí the PRI gubernatorial candidate took office but was shortly removed by the president, who replaced him with a national PRI leader as the interim governor. In Guanajuato the president removed the PRI candidate before he was installed, substituting a mayor from the PAN. In 1992 state leadership claimed victory in elections in Tabasco; this too eventuated in the removal of the governor.

Each of the three elections had something in common: the opposition planned a protest march to Mexico City, objecting to the alleged fraud. The president of Mexico intervened in all three states, forcing the governor or governor-elect from office. In fact, by 1992 Salinas had removed eight governors, more than any other president in forty years,[24] which illustrates two important patterns in his administration. First, the president made the decisions, and the president extralegally brought about resolution of the disputes. Second, Salinas legitimized opposition protests as a viable means of expressing demands. Each time he reacted by removing a governor, he encouraged protests in the future.

A certain irony exists in Salinas's liberalization philosophy. Economically, he has tried to illustrate that Mexico is decentralizing, allocating decision-making authority to numerous independent enterprises. Politically, although he asserted he desired liberalization for his own party leadership and in national politics, the results demonstrate otherwise: as the government decentralizes economic decision making, it actually furthers centralized political decision making.

Other characteristics of Salinas's political strategy illustrate the complexity of his policies and the contradictions in his goals. The most radical political policy of his administration involved long-standing Church-state relations. Early on he announced his desire to "modernize" Church-state relations. He invited Church leaders to his inauguration, an unprecedented gesture, and followed this by appointing a personal representative to the Vatican.

In December 1991 Salinas proposed changes in the Constitution governing Church-state relations. Among the changes later approved by the Chamber of Deputies, were recognizing all churches as legal entities; allowing priests and ministers to vote and to run for political office if they had resigned from their clerical offices five years earlier; permitting churches and religious officials to offer primary and secondary education as long as they respected government-approved plans of study; allowing public celebrations of religious ceremonies; granting churches the right to own property; and explicitly separating churches from the state.

Although clergy traditionally violated some constitutional provisions in practice, the president confronted a highly emotional issue with deep roots in Mexican political liberalism. The proposals, however, not only modernized the relationship between religion and the state but they conformed to explicit principles of democracy, particularly concerning the right to vote. On the other hand, Mexico's general record on human rights has contradicted many of the goals implicit in democratization.

Again, as in the case of economic liberalization, the United States has been an important, if indirect, actor in Mexico's internal political affairs. In testimony before the House Committee on Foreign Affairs, leaders of international human rights organizations described many human rights abuses. Specifically, Amnesty International's representative, the deputy director of its Washington, D.C., office, stated, "Mexico is a country with staggering levels of political violence."[25]

Thus, not only were human rights abuses in the criminal justice system identified as endemic in many international and domestic reports, but some government officials intimidated political opponents and independent observers connected with the media. Reporters were beaten, kidnapped, threatened, pressured, and deported for articles critical of the president, his family, and his policies. A closed, not an open, climate for public discussion of important issues exists in Mexico.[26]

International pressure forced Salinas to establish the National Commission for Human Rights, presided over by Jorge Carpizo, a highly respected jurist and Supreme Court justice.[27] The president's primary motivation was to temper worldwide disapproval of Mexico's human rights record in anticipation of public discussion and congressional hearings in the United States on the proposed free trade agreement. The commission has been able to investigate some complaints and protect the rights of some abused citizens, but the government's commitment to full investigation and vigorous prosecution is questionable. The most notorious example of the Mexican government's commitment occurred in the late fall of 1991. In a confrontation between Mexican army troops and agents of the federal attorney general's office, the troops, protecting a landing strip in Veracruz for drug traffickers killed seven agents. A chase plane pursued the drug plane. U.S. drug enforcement agents filmed the encounter. The president gave the Commission for Human Rights Commission responsibility for investigating. Although it issued a 104-page report, and the army tried and convicted the zone commander and other officers, neither the report nor the trial gave an explanation for the murders.[28]

National, state, and local election results in 1990, 1991, and 1992, presidential intervention in electoral disputes, and failure to respond fully to human rights abuses reveal the slow pace and incompleteness of political reform and democratization in Mexico. Electoral integrity is not the only test of the government's commitment to democracy. The public's cynicism about government institutions is also a consequence of pervasive corruption in public life. It is not just obvious forms of corruption that detract from the integrity of the institutions and the leadership but corruption's detrimental effects on the sense of community interest and trust. Thus, pervasive corruption complements the belief that public life provides opportunities to benefit family and friends rather than serve the interests of all

Mexicans. This is not to say that all public officials in Mexico are dishonest but to suggest than an ambience favorable to self-interest, dishonesty, and favoritism prevails in many areas and at numerous levels of the public sector. The Mexican public itself deems corruption widespread.[29]

MEXICO'S FUTURE

Mexico today faces a democratization challenge to its semiauthoritarian political model emanating from its citizenry and from dissident and opposition elites. The challenge is formidable, and has many consequences for the leadership in the short and long term. It is also clear that the economic strategy pursued by the government, especially since 1988, is affecting and will be affected by the trends in political liberalization.

If Mexico is undergoing a slow, tortuous process of democratization, what exactly does that imply? As several analysts have suggested, democracy incorporates the following: policy debates and political competition, citizen participation, accountability of the rules to law and representative mechanisms, civilian control over the military, and respect for the views and rights of others.[30]

It is not enough to introduce the mechanisms or processes of democracy; in this respect, Mexico has already achieved certain facets of a democratic system. What is equally important, and more difficult to accomplish, is a certain level of societal trust, which in turn encourages individuals to respect the political views of others.[31]

The Mexican government has chosen to pursue economic and political liberalization. Although Salinas has argued for the necessity of economic liberalization first, followed by a more evolutionary political liberalization, no guarantees exist that economic decentralization and foreign investment will automatically produce democratic political tendencies. There is no doubt that the two trends are intertwined, and that certain aspects of Mexico's economic liberalization strategy will enhance political developments, including decentralization of economic authority, competition, the introduction of international influences, and the exchange of ideas.[32] Most observers believe, for example, that if Canada, Mexico, and the United States achieve a free trade agreement, it will have certain consequences for democratization and yield benefits thereto.[33]

To what degree does Mexico promote policy debate and political competition? Since the 1980s political competition has greatly increased, measured in terms of opposition successes in the voting booths. Competition reached a high point in the 1988 presidential election. Although its meaning is confused because many citizens voted against the PRI rather than for a specific alternative, the fact is that half the voters, according to the official statistics, voted for a nongovernment candidate, and in reality probably more than half voted for other parties. Three years later the 1991 elections revealed an entirely different situation; opposition strength appeared to have vanished in thin air. Despite allegations of widespread fraud by opposition parties, empirical survey research pointed to a revival of PRI strength and appeal while support for the newest party, the PRD, waned. Nevertheless, the desire of the government party to overwhelm its opponents and to claim a sweeping victory negated any interpretation that viable political competition existed. As one critic suggested, Mexico created electoral pluralism *without* competition.[34]

Electoral contests have always received most of the attention in the media and in analyses of political competition. Perhaps the most obvious weakness in the measuring of Mexican democratic achievements is in the policy arena. The argument in the past has been that political leadership and the governing party representing that leadership were sufficiently broad, ideologically speaking, to incorporate opposing points of view. Although policy debates rarely spilled over into public view, considerable give-and-take occurred within the top executive-branch echelon. In recent years, however, although this pattern was still the norm, Salinas narrowed the range of acceptable policy alternatives, especially in the economic realm, forcing dissident members of his own party from leadership posts.

From 1988 to 1991, for the first time in recent Mexican history, the opposition acquired sufficient strength in the Chamber of Deputies to give the lower house a strong voice in the policy process—especially when the government desired constitutional amendments, for which it lacked a sufficient proportion of PRI legislative votes. Debate over significant policy issues opened up when the government party sought allies to support constitutional legislation. The 1991 elections damped this departure, reducing public debate. Even within its own ranks, the government maintained very strict discipline, not permitting officials to speak out on controversial policy issues. This is illustrated with the prospective free trade agreement with the United States. Severe media critics, especially those who focus on Salinas personally, have endured numerous trials, including threats of physical harm to themselves and their families. In short, the government has not promoted a climate favorable to policy debate but the opposite. It has reversed the more congenial intellectual environment of the de la Madrid administration.

Citizen participation has undergone several changes in recent years.

The 1988 presidential contest generated considerable interest and hence participation. It is difficult, however, to measure citizen involvement on the basis of willingness to cast ballots because of questionable election data. In the recent past participation declined as cynicism toward the electoral process burgeoned. Still, it would be fair to say that the development of a viable third party reflected new levels of citizen participation. Perhaps more important, grass-roots organizations flourished. They encompass a broad spectrum of citizens and interests. These are positive signs, yet the ability of citizens to participate in the decision-making process or to have their interests directly represented at policy-making levels is minimal. This is due in part to the structure of the relationship between the legislative and executive branches, and the dominance of the latter over the policy process.

To most Mexicans, the accountability of their leadership is a moot point. Although individual citizens can and do use the legal system to protect themselves from abuses by the executive branch, including the presidency, they are few in number. Actions of the executive branch, even when carrying out policy decisions favored by most Mexicans, often contradict legal guidelines or ignore institutional channels. Many of Salinas's decisions have not been legally implemented, whether they involved criminal matters, such as arresting a drug dealer, or were purely political, like his removals of state governors.

The continued use of the presidency to resolve political questions when legal channels are at hand, has numerous consequences. It puts the presidency above all other institutions, giving it extraordinary powers in practice and in the eyes of the citizenry. Hence, when the president exercises authority extralegally, he creates expectations on the part of the average Mexican. Even when the president intervenes to overturn elections and to support politically liberalizing trends, the intervention itself contravenes these goals. The justification that the ends justifies the means is contradictory to democratization.

The presidency is the most important institutional force in Mexican society, and as such, serves as the leading model for other forms of political and social behavior. If the presidency behaves in a paternalistic, interventionist, and authoritarian manner, it sets the tone for the rest of the political system and society.

The rule of law has yet to develop substantial prestige in Mexico. Analysts argue that the influx of foreign businesses and greater international competitiveness will encourage heavier reliance on the legal system to resolve disputes, given that these new phenomena traditionally used similar channels in their own cultures. This may well be the case, but for the average Mexican, the law provides few guarantees of equal treatment. As the human rights literature make clear, the criminal justice system is often the abuser of human rights, not their protector. The level of corruption also influences the legal system, and the law's applicability to all citizens.

For much of Latin America, and many other Third World countries, civilian control over the military has rarely existed. The Mexican civilian sector, however, has had an exceedingly successful relationship with the military, especially since the 1930s, and has unquestionably maintained its supremacy. Even so, the relationship is quite complex and involves many variables. To achieve civilian supremacy, civilian leadership gave up a part of its autonomy regarding military matters, including promotions and internal allocation of funding. As the event in Veracruz illustrated, civilian leadership, including the president, put self-imposed limits on their ability to discipline errant and corrupt behavior by members of the armed forces. Civilian leadership retained its position in part by never publicly criticizing the armed forces, actually by heaping praise on its institutional integrity even when that may not have been the case.[35]

Finally, as our earlier analysis of citizen political values and attitudes illustrated, many Mexicans share beliefs conducive to democratic behavior. However, large portions of the populace and of the leadership, opposition and establishment alike, have not yet accepted the obligations of a democratic culture, one of which is tolerance of opposing political and social views. Establishment leadership's positive actions toward its own members and toward members of the opposition can encourage this value. The degree to which the electoral process remains open, and highly competitive also encourages this value.

Mexico's political development has taken some interesting paths in the past decade. Its politicians and political process labor under a special burden with which few countries have had to cope: proximity to the United States. As noted in the discussion of the historical evolution of Mexico's political characteristics, just the geographic nearness of the United States and its historic involvement in internal Mexican affairs, flavored earlier eras.

The United States continues to exercise considerable influence on Mexico's leadership. The influence is typically implicit and indirect. In fact, if the United States were to attempt to influence Mexican political affairs directly, the effort would surely backfire. Nevertheless, the influence is obvious and many actors are involved. For example, U.S. media often criticize Mexico for its political failures, focusing in recent years on electoral fraud, corruption, and human rights abuses. Mexicans often resort to nationalism to deny the veracity and importance of the criticism. On the other hand, opposition leaders cite the reports to justify their own criticism and to take the leadership to task. Also, in a search for greater legitimacy, establishment leadership and the Salinas administration particularly use favorable reports in the international press to gain credibility within Mexico and abroad.

The United States has exercised an indirect influence since the 1980s on many aspects of government policy through its drug interdiction program. Drug enforcement agents have worked closely with their Mexican counterparts to stanch the northward flow of drugs. The involvement of U.S. representatives in matters having to do with Mexican national security often raises important political issues. It is very unlikely that the confrontation between the army and agents of the attorney general's office in late 1991 would have come to the attention of the National Commission for Human Rights, the Secretariat of national defense, and the president if agents of the U.S. Drug Enforcement Agency had not been present.

The posture of the U.S. government toward various Mexican policies can flavor the political environment. This is not to suggest that the United States alone can or does determine political decisions in Mexico, but it may well bear on their outcome. For example, until 1988 some U.S. officials were unceasingly critical of Mexico's failure to increase electoral competitiveness, and to provide honest vote counts. Some observers noted, however, that after 1988 such criticism was tempered or dropped altogether. The change was motivated, it was suggested, by U.S. support for Salinas's economic strategy and by the perception that the populist Left, which opposes Salinas's economic strategy, would be the primary beneficiaries of a political opening.[36]

Whether or not Mexico's leadership is truly committed to democratic reforms, the fact is that many forces within and outside Mexico are pushing it in the direction of greater democracy, with all that entails.[37] Although the contradictions between strong state-led economic liberalization and political democracy abound, it is not likely that the leadership can sustain the separation indefinitely. Democratization is slowly eating away at the framework of authoritarianism in Mexico, creating the hybrid political model of the future. Whatever path it takes, Mexico will carry its own special politics into the twenty-first century.

NOTES

1. Pedro Aspe, "Thoughts on the Structural Transformation in Mexico: The Case of Privatization of Public Sector Enterprises" (Speech to the World Affairs Council, Los Angeles, June 21, 1991), 6.

2. Samuel Schmidt, *The Deterioration of the Mexican Presidency: The Years of Luis Echeverría* (Tucson: University of Arizona Press, 1991).

3. Luis Rubio and Roberto Newell, *Mexico's Dilemma: The Political Origins of Economic Crisis* (Boulder, Colo.: Westview Press, 1984), 134.

4. George Grayson, *The Politics of Mexican Oil* (Pittsburgh: University of Pittsburgh Press, 1980), 119.

5. Sidney Weintraub, *A Marriage of Convenience: Relations between Mexico and the United States* (New York: Oxford University Press, 1990), 134ff.

6. Sylvia Maxfield, "Introduction," in *Government and Private Sector in Contemporary Mexico*, ed. Sylvia Maxfield and Ricardo Anzaldúa Montoya (La Jolla: Center for U.S.-Mexican Studies, UCSD, 1987), 18.

7. Daniel Levy and Gabriel Székely, *Mexico: Paradoxes of Stability and Change* (Boulder, Colo.: Westview Press, 1987), 157.

8. Wayne A. Cornelius, "The Political Economy of Mexico under de la Madrid: Austerity, Routinized Crisis, and Nascent Recovery," *Mexican Studies/Estudios Mexicanos* 1 (Winter 1985): 83–124.

9. Personal interview, Mexico City, July 20, 1984.

10. Peter H. Smith, "The 1988 Presidential Succession in Historical Perspective," in *Mexico's Alternative Political Futures*, ed. Wayne A. Cornelius, Judith Gentleman, and Peter H. Smith (La Jolla: Center for U.S.-Mexican Studies, 1989), 402.

11. Tom Barry, ed., *Mexico: A Country Guide* (Albuquerque: Inter-Hemispheric Education Resource Guide, 1992), 86.

12. Marjorie Miller and Juanita Darling, "Progress and Promise," *Los Angeles Times*, October 22, 1991.

13. *Mexico Report*, February 10, 1992, 6.

14. Ibid., 6.

15. Raymundo Riva Palacio, "Mexico Is Not an Island," *El Financiero International*, February 24, 1992, 17; Sergio Sarmiento, "Solidarity Offers Hope for Votes," ibid., September 30, 1991, 12.

16. "A New Hope for the Hemisphere," *New Perspective Quarterly*, 8 (Winter 1991): 128.

17. For some unusual insights into his strategy, see Robert A. Pastor, "Post-Revolutionary Mexico: The Salinas Opening," *Journal of Inter-American Studies and World Affairs* 32 (Fall 1990): 1–22.

18. For a discussion of this issue, see Michael Coppedge, "Mexican Democracy: You Can't Get There from Here," in *The Politics of Economic Liberalization in Mexico*, ed. Riordan Roett (Boulder, Colo.: Lynne Rienner, 1993).

19. John J Bailey, "Mexico," in *Latin American and Caribbean Contemporary Record, 1985–86*, ed. Jack W. Hopkins (New York: Holmes & Meier, 1987), B355–58; Wayne A. Cornelius, "Political Liberalization in an Authoritarian Regime: Mexico, 1976–1985," in *Mexican Politics in Transition*, ed. Judith Gentleman (Boulder, Colo.: Westview Press, 1987), 15–40.

20. For various analyses of the results, see Edgar W. Butler and Jorge A.

Bustamante, eds., *Sucesión Presidencial: The 1988 Mexican Presidential Election* (Boulder, Colo.: Westview Press, 1991).

21. Roderic A. Camp, "Mexico's 1988 Elections: A Turning Point for Its Political Development and Foreign Relations," in *Sucesión Presidencial: The 1988 Mexican Presidential Elections,* ed. Edgar W. Butler and Jorge A. Bustamante (Boulder, Colo.: Westview Press, 1990), 104–8.

22. For interesting insights into the issue of leadership, from the point of view of a Mexican intellectual, see Federico Reyes Heroles, "De la debilidad al liderazgo," *Este País,* September 1991, 3–10.

23. Robert R. Bezdek, "Electoral Opposition in Mexico: Emergence, Suppression, and Impact on Political Processes" (Ph.D. diss., Ohio State University, 1973).

24. Ted Bardacke, "Another Governor Bites the Dust," *El Financiero International,* February 10, 1992, 13.

25. House Committee on Foreign Affairs, *Hearing before the Subcommittee on Human Rights and International Organizations, and on Western Hemisphere Affairs, September 12, 1990* (Washington, D.C.: GPO, 1990), 31.

26. For a short list, see Andrew Reding and Christopher Whalen, "Fragile Stability, Reform and Repression in Mexico under Carlos Salinas, 1989–1991" (New York: World Policy Institute, New School for Social Research, 1991), 12ff.

27. See National Commission for Human Rights, *Third Report, June–December, 1991* (Mexico, 1991), for details.

28. *Washington Post,* January 14, 1992; *Los Angeles Times,* January 15, 1992; ibid., December 7, 1991.

29. Stephen D. Morris, *Corruption and Politics in Contemporary Mexico* (Tuscaloosa: University of Alabama Press, 1991), 103. Morris provides the most detailed analysis of corruption and politics in print.

30. See Terry Lynn Karl, "Dilemmas of Democratization in Latin America," *Comparative Politics* 23 (October 1990): 2–3; David Llehmann, *Democracy and Development in Latin America: Economics, Politics and Religion in the Postwar Period* (Philadelphia: Temple University Press, 1990), 206.

31. For arguments in support of this interpretation, see my "Political Liberalization: The Last Key to Economic Modernization in Mexico," in *The Politics of Economic Liberalization in Mexico,* ed. Riordan Roett (Boulder, Colo.: Lynne Rienner, 1993).

32. See Luis Rubio, "Economic Reform and Political Liberalization," in *The Politics of Economic Liberalization in Mexico,* ed. Riordan Roett (Boulder, Colo.: Lynne Rienner, 1993).

33. For an excellent presentation of possible alternative scenarios involving the impact of free trade on Mexican politics, see Peter H. Smith, "The Political Impact of Free Trade on Mexico" (Paper presented at symposium, The Social, Political, and Cultural Implications of a North American Free Trade Area, El Colegio de México, March 1991.

34. Juan Molinar Horcasitas, *El tiempo de la legitimidad, elecciones, autoritarismo y democracia en México* (México: Cal y Arenas, 1991), 247.

35. For details about this relationship, see my chapter "What Kind of Relationship?" in *Generals in the Palacio: The Military in Modern Mexico* (New York: Oxford University Press, 1992), 212ff.

36. For a sophisticated presentation of this argument, see Lorenzo Meyer, "Mexico: The Exception and the Rule," in *Exporting Democracy: The United States and Latin America,* ed. Abraham F. Lowenthal (Baltimore: Johns Hopkins University Press, 1991), 227.

37. For an excellent overview of what has occurred and the dilemmas it poses for Mexico, see John J Bailey and Leopoldo Gómez, "The PRI and Political Liberalization," *Journal of International Affairs* 43 (Winter 1990): 291–312; Adolfo Gilly, "The Mexican Regime in Its Dilemma," *Journal of International Affairs* 43 (Winter 1990): 273–90.

Bibliographic Essay

For the student initially exploring Mexican politics, a voluminous literature exists in both Spanish and English. The purpose of this essay is only to suggest sources that can lead to more detailed analyses. The focus is largely on those readily available, generally in English, and published in the past decade.

An excellent place to start bibliographically for Latin American politics and Mexican politics specifically is David W. Dent, *Handbook of Political Science Research on Latin America: Trends from the 1960s to the 1990s* (Westport, Conn.: Greenwood Press, 1990), which includes chapters on various countries, several thematic topics, and two chapters on Mexico, covering domestic and international affairs respectively. For a more complete survey of recent material on Mexico, in all languages, with brief annotations, see my "Government and Politics, Mexico," in *Handbook of Latin American Studies: Social Sciences,* vol. 51, ed. Dolores Moyano Martin (Austin: University of Texas Press, 1991), 474–485, as well as previous volumes.

Some excellent broad surveys of various facets of Mexican politics, decision making, and political economy exist. An initial source is John J Bailey, *Governing Mexico: The Statecraft of Crisis Management* (New York: St. Martin's Press, 1988), which examines the administrative and structural capabilities of the party and the state within the context of recent reforms. For a broad political, social, and economic overview, the best general survey is Daniel Levy and Gabriel Székely, *Mexico: Paradoxes of Stability and Change,* 2d ed. (Boulder, Colo.: Westview Press, 1987), which tends to focus on Mexico's relationship to the United States, providing a joint Mexican and North American interpretation. A Mexican view, from a political-economy perspective, can be found in Roberto Newell and Luis Rubio, *Mexico's Dilemma: The Political Origins of Economic Crisis* (Boulder, Colo.: Westview Press, 1984). Another valuable interpretation, particularly of the Luis Echeverría (1970–1976) and José López Portillo (1976–1982) periods can be found in Judith Hellman, *Mexico in Crisis,* 2d ed. (New York: Holmes & Meier, 1983). On the presidency itself, the best work is George Philip, *The Presidency in Mexican Politics* (New York: St. Martin's Press, 1992). For a recent Mexican perspective, see Miguel Basáñez, *La lucha por la hegemonia en México, 1968–1990,* 8th ed. (México: Siglo XXI, 1990).

Several edited collections provide up-to-date interpretations of developments immediately before the 1988 presidential election, and consequences since 1988.

These include Judith Gentleman, ed., *Mexican Politics in Transition* (Boulder, Colo.: Westview Press, 1987); and Wayne A. Cornelius, Judith Gentleman, and Peter H. Smith, eds., *Mexico's Alternative Political Futures* (La Jolla: UCSD Center for U.S.-Mexican Studies, 1989). Both of these works contain some of the best interpretative essays on recent Mexico by North American and Mexican scholars.

The only extensive comparative analysis of Mexico and another country is that of Brazil and Mexico, found in Sylvia Ann Hewlett and Richard S. Weinert's excellent *Brazil and Mexico: Patterns in Late Development* (New York: ISHI, 1982). Of course, various edited collections include Mexico as part of a larger regional approach. These often are not well integrated. Among the best is Daniel Levy, "Mexico: Sustained Civilian Rule without Democracy," in *Democracy in Developing Countries: Latin America,* vol. 4, ed. Larry Diamond, Juan J. Linz, and Seymour Martin Lipset (Boulder: Lynne Rienner, 1989), 459–97.

The historical literature on Mexico is extensive and detailed, although gaps exist for many topics and periods. This topic alone would consume more space than is available here. The best comprehensive survey of Mexican history is Michael Meyer and William B. Sherman, *The Course of Mexican History,* 4th ed. (New York: Oxford University Press, 1991), which details readings for various periods and subjects, many of them essential classics. For a broad overview of political characteristics stemming from the colonial system, see Colin M. MacLachlan, *Spain's Empire in the New World* (Berkeley: University of California Press, 1991). For the late-nineteenth-century political heritage, John H. Coatsworth, "Los origines del autoritarismo moderno en México," *Foro Internacional* 16 (October–December 1975): 205–32; and Charles A. Hale, *The Transformation of Liberalism in Late Nineteenth Century Mexico* (Princeton: Princeton University Press, 1989), provide helpful interpretations. From a Mexican point of view, especially for understanding political development and the importance of culture in politics, it is helpful to read Justo Sierra, *The Political Evolution of the Mexican People* (Austin: University of Texas Press, 1969); Samuel Ramos, *Profile of Man and Culture in Mexico* (Austin: University of Texas Press, 1962); and Octavio Paz, *The Labyrinth of Solitude: Life and Thought in Mexico* (New York: Grove Press, 1961).

A broad understanding of the theoretical debates concerning Mexico's unique political model is found in a revealing discussion in the now-classic "Who Rules Mexico? A Critique of Some Current Views of the Mexican Political Process," by Carolyn and Martin Needleman in *Journal of Politics* 31 (November 1969): 1011–34. For a variety of interpretations focusing on authoritarianism in Mexico, see José Luis Reyna and Richard Weinert, eds., *Authoritarianism in Mexico* (New York: ISHI, 1977). An interpretation stressing the corporatist flavor of the regime is lucidly conveyed by Ruth Spalding, "State Power and Its Limits: Corporatism in Mexico," *Comparative Political Studies,* 14 (July 1981), 139–64; and John W. Sloan, "The Mexican Variant of Corporatism," *Inter-American Economic Affairs* 38 (Spring 1985): 3–18.

The subject of contemporary Mexican political culture has received more attention from Mexicans than from Americans. The comparative classic on this topic, although flawed in places, is Gabriel Almond and Sidney Verba, *The Civic Culture: Political Attitudes and Democracy in Five Nations* (Boston: Little, Brown, 1965), and the even more useful follow-up analysis in their edited work, *The Civic Culture Revisited* (Boston: Little, Brown, 1980). Another work, again comparing Brazil and Mexico, although having less to do with political values, is that of Joseph A. Kahl, *The Measurement of Modernism: A Study of Values in Brazil and Mexico* (Austin: University of Texas Press, 1974). The most recent comparative study, although focusing on international linkages, is that by Ronald Inglehart, Neil Nevitte, and Miguel Basáñez, *North American Convergence* (Princeton: Princeton University Press, forthcoming 1993). Several excellent works have appeared in Mexico that provide insightful data on changing Mexican values socially and politically. The two most comprehensive surveys are Alberto Hernández Medina and Luis Narro Rodríguez, eds., *Cómo somos los mexicanos* (México: CREA, 1987), and Enrique Alduncin, *Los valores de los mexicanos* (México: Fomento Cultural Banamex, 1986), and a more recent survey, *Los valores de los mexicanos, México en tiempos de cambio,* 2 (Mexico: Fomento Cultural Banamex, 1991).

On the relationship between cultural values and democracy, a good place to begin is Ronald Inglehart, "The Renaissance of Political Culture," *American Political Science Review* 82 (November 1988): 1219. For some useful comparisons with Mexico, see Mitchell Seligson, "Political Culture and Democratization in Latin America," in *Latin American and Caribbean Contemporary Record,* ed. James Malloy and Eduardo A. Gamarra (New York: Holmes & Meier, 1990), A49–65. For Mexico, the best analysis is that of John Booth and Mitchell Seligson, "The Political Culture of Authoritarianism in Mexico," *Latin American Research Review,* 19, no. 1 (1984): 106–24.

Moving from political culture to the more specific issue of how values affect partisanship, alienation, and tolerance, we discover very little literature on Mexico. One major study is that by Rafael Segovia, *La politización del niño mexicano* (México: El Colegio de México, 1975), which analyzes data from Mexico City schoolchildren, offering many useful comparisons that can be drawn from similar studies of the United States. Other than the Almond and Verba book, the most important analysis, a case study, is that by Richard Fagen and William Tuohy, *Politics and Privilege in a Mexican City* (Stanford: Stanford University Press, 1972), which explores citizen attitudes in the Veracruz capital Jalapa. The best general analysis of this subject is still Ann Craig and Wayne Cornelius, "Political Culture in Mexico: Continuities and Revisionist Interpretations," in *The Civic Culture Revisited,* ed. Gabriel Almond and Sidney Verba (Boston: Little, Brown, 1980) 325–93.

For specific groups or issues tied to political behavior, few studies are available in English or Spanish. On partisanship, one of the best works is Joseph Klesner, "Changing Patterns of Electoral Participation and Official Party Support in Mexico," in *Mexican Politics in Transition,* ed. Judith Gentleman (Boulder, Colo.: Westview Press, 1987), 95–127. For the impact of religion, see my "Religion and Politics: The Laity in Mexico" (unpublished paper); and Charles L.

Davis, "Religion and Partisan Loyalty: The Case of Catholic Workers in Mexico," *Western Political Quarterly* 45 (March 1992): 275–97. On the subject of gender, the only analysis specifically on this topic is William J. Blough, "Political Attitudes of Mexican Women: Support for the Political System among a Newly Enfranchised Group," *Journal of Inter-American Studies and World Affairs* 14 (May 1972): 201–24.

Two general works are available on political recruitment. They are Peter H. Smith, *Labyrinths of Power: Political Recruitment in Twentieth-Century Mexico* (Princeton: Princeton University Press, 1979), which explores these patterns from 1900 through the Echeverría administration; and Roderic A. Camp, *Education and Recruitment in Twentieth Century Mexico* (Tucson: University of Arizona Press, 1980), which analyzes the linkage between higher education and public life as a locus of recruitment. For a more recent analysis of some trends, see Guillermo Kelley, "Politics and Administration in Mexico: Recruitment and Promotion of the Politico-Administrative Class," Technical Papers Series, No. 33 (Austin: ILAS, University of Texas, 1982); Roderic A. Camp, "The Political Technocrat in Mexico and the Survival of the Political System," *Latin American Research Review* 20, no. 1 (1985): 97–118; and my "Camarillas in Mexican Politics: The Case of the Salinas Cabinet," *Mexican Studies* 6 (Winter 1990): 85–108. For an understanding of the role of informal groups and upwardly mobile political careers, see Merilee S. Grindle, "Patrons and Clients in the Bureaucracy: Career Networks in Mexico," *Latin American Research Review* 12, no. 1 (1977): 37–66; and Luis Roniger, *Hierarchy and Trust in Modern Mexico and Brazil* (New York: Praeger, 1990).

An area in domestic politics that has received considerable attention, theoretically and substantively, has been the relationship between the government and important interest sectors. The theory of corporatism, which has been most often used to analyze this relationship in Mexico, has been well described in the articles mentioned above by Sloan and Spalding. For a useful historical view set earlier in the twentieth century, see Nora Hamilton, *The Limits of State Autonomy: Post Revolutionary Mexico* (Princeton: Princeton University Press, 1982). For an interpretation of the changes wrought in the late 1980s, see Howard J. Wiarda, "Mexico: The Unravelling of a Corporatist Regime?" *Journal of Inter-American Studies and World Affairs* 30 (Winter 1988–1989): 1–28.

To understand policy-making generally in Mexico, the best sources include John Bailey, *Governing Mexico*, mentioned above, and Judith A. Teichman, *Policymaking in Mexico: From Boom to Crisis* (Boston: Allen & Unwin, 1988). For specific case studies dealing with the bureaucracy, see Susan Kaufman Purcell, *The Mexican Profit-Sharing Decision: Politics in an Authoritarian Regime* (Berkeley: University of California Press, 1975); Peter Ward, *Welfare Politics in Mexico: Papering over the Cracks* (London: Allen & Unwin, 1986); and Merilee Grindle, *Bureaucrats, Politicians, and Peasants in Mexico: A Case Study in Public Policy* (Berkeley: University of California Press, 1977).

For specific groups and their relationship to the state, several studies are available. In the context of the relationship to the military, see Roderic Ai Camp, *Generals and the Palacio: The Military in Modern Mexico* (New York: Oxford

University Press, 1992), especially the final chapter, for a broad overview from the 1940s to the present; Michael J. Dziedzic, "The Essence of Decision in a Hegemonic Regime: The Case of Mexico's Acquisition of a Supersonic Fighter" (Ph.D. diss., University of Texas, Austin, 1986), an excellent case study of Mexico's decision to acquire fighter planes from the United States; Phyllis Greene Walker, "The Modern Mexican Military: Political Influence and Institutional Interests" (M.A. thesis, American University, 1987), based on numerous interviews, which explores the topic generally; and David Ronfeldt, ed., *The Modern Mexican Military: A Reassessment* (La Jolla: UCSD Center for U.S.-Mexican Studies, 1984), a collection by leading students of the military.

For business and politics, the best work on industrial groups is that by Dale Story, *Industry, the State, and Public Policy in Mexico* (Austin: University of Texas Press, 1986). For the broad relationship between businesspeople and the state, see my *Entrepreneurs and the State in Twentieth Century Mexico* (New York: Oxford University Press, 1989). An excellent case study, although having important theoretical implications, is Douglas C. Bennet and Kenneth E. Sharpe, *Transnational Corporations versus the State: The Political Economy of the Mexican Auto Industry* (Princeton: Princeton University Press, 1985). A broad collection, incorporating recent Mexican and North American scholarship is Sylvia Maxfield, ed., *Government and the Private Sector in Contemporary Mexico* (La Jolla: UCSD Center for U.S.-Mexican Studies, 1987).

The most neglected link in this relationship is that between the Catholic Church and state. The best recent comprehensive work is Roberto Blancarte, *El poder salinismo e Iglesia católica, una nueva convivencia?* (México: Grijalbo, 1991). Useful descriptions of this relationship in English are Karl Schmitt, "Church and State in Mexico: A Corporatist Relationship," *Americas* 40 (January 1984): 349–76, which provides a succinct but clear historical overview; and Claude Pomerlau, "The Changing Church in Mexico and Its Challenge to the State," *Review of Politics,* 43 (October 1981): 540–59.

Labor's relationship to the Mexican government has received more attention than the Church or the military. The best general interpretation on this topic can be found in Kevin J. Middlebrook, ed., *Unions, Workers, and the State in Mexico* (La Jolla: UCSD Center for U.S.-Mexican Studies, 1991), especially his introductory chapter, "State-Labor Relations in Mexico: The Changing Economic and Political Context," 1–26. Recent labor relations under Salinas are described in Dan LaBotz, *Mask of Democracy: Labor Suppression in Mexico Today* (Boston: South End Press, 1992). A short but helpful overview is also provided in George Grayson, *The Mexican Labor Machine: Power, Politics, and Patronage,* Significant Issues Series (Washington, D.C.: CSIS, 1989).

The other group whose relationship is analyzed is intellectuals, a more amorphous sector institutionally. Various studies explore the development and contributions of individual groups of intellectuals, especially writers, but the literature is sparse on the issue of intellectual-state relationships broadly speaking. For this general relationship, see Roderic A. Camp, *Intellectuals and the State in Twentieth Century Mexico* (Austin: University of Texas Press, 1985). For an important analysis of dissent in Mexico, see Evelyn P. Stevens, *Protest and Response in Mexico* (Cambridge: MIT Press, 1974). This subject is also dealt with by Judith Hellman, the Levy and Székely work, and Basáñez's book.

The branches of government, and their relationship to the policy-making process, have been neglected completely in the literature on Mexican politics. For the legislative branch, the only comprehensive study is that by Rudolfo de la Garza, "The Mexican Chamber of Deputies and the Mexican Political System" (Ph.D. diss., University of Arizona, 1972), which of course describes features no longer present in this branch. Richard Bath, "The Mexican Congress: A New Role?" *Proceedings from the 37th Annual Meeting of the RMCLAS* (February 1989, Las Cruces, New Mexico), 82–90, explores the changing relationship of Congress to the executive branch. Although not dealing with the legal system specifically, George A. Armstrong, *Law and Market Society in Mexico* (New York: Praeger, 1989), discusses the evolution of the legal system in relation to capitalism.

The government party has received more attention than the branches of government. The best recent analysis in English is Dale Story, *The Mexican Ruling Party: Stability and Authority* (Stanford: Hoover Institution, 1986). The most comprehensive historical analysis of the party is Javier Luis Garrido, *El partido de la revolución institucionalizada, la formación del nuevo estado en México (1928–1945)* (México: Siglo XXI, 1982).

Opposition parties continue to attract considerable attention in the literature but typically within an electoral context. For the National Action Party, the best institutional analysis is Donald Mabry, *Mexico's Acción Nacional: A Catholic Alternative to Revolution* (Syracuse: Syracuse University Press, 1973). Abraham Nuncio, *El PAN* (Mexico: Editorial Nueva Imagen, 1986), provides some insights about recent PAN activities, but Soledad Loaeza, "Derecha y democracia en el cambio político mexicano, 1982–1988," *Foro Internacional* 30 (April–June 1990), does it more objectively. The PRD has not received adequate treatment because of its recent origins. However, for some views of the Democratic Revolutionary Party through its leadership, see Jesús Galindo López, "A Conversation with Cuauhtémoc Cárdenas," *Journal of International Affairs* 43 (Winter 1990): 395–406; and Carlos B. Gil, *Hope and Frustration: Interviews with Leaders of Mexico's Political Opposition* (Wilmington, Del.: Scholarly Resources, 1992). Miguel Angel Centeno's short *Mexico in the 1990s: Government and Opposition Speak Out*, Current Issue Brief (La Jolla: UCSD Center for U.S.-Mexican Studies, 1991), provides some useful insights into differences in their joint views on policy issues.

Analyses of elections, election data, and the 1988 election specifically abound. Some of the better work in English can be found in Gentleman, *Mexican Politics in Transition;* Cornelius et al., *Mexico's Alternative Political Futures;* and Edgar W. Butler and Jorge A. Bustamante, *Sucesión Presidencial: The 1988 Mexican Presidential Election* (Boulder, Colo.: Westview Press, 1991), the most detailed examination in English of this benchmark election. A more comprehensive analysis leading up to 1988 is Joseph A. Klesner, "Electoral Reform in an Authoritarian Regime: The Case of Mexico" (Ph.D. diss., MIT, February 1988), and Arturo Alvarado Mendoza, ed., *Electoral Patterns and Perspectives in Mexico* (La

Jolla: UCSD Center for U.S.-Mexico Studies, 1987), contains Mexican and North American interpretations. For a post-1988 analysis, see Keith Yanner, "Democratization in Mexico, 1988–1992" (Ph.D. diss., Washington University, 1992), and Juan Molinar Horcasitas, *El tiempo de la legitimidad, elecciones, autoritarismo y democracia en México* (México: Cal y Arena, 1991).

Political and economic liberalization has received considerable attention since 1987, especially after Salinas's first year in office. A broad theoretical context can be found in Terry Lynn Karl, "Dilemmas of Democratization in Latin America," *Comparative Politics* 23 (October 1990): 1–21. For highly critical views of Salinas's political failures, see Andrew Reding's essays, for example, "Mexico under Salinas: A Facade of Reform," *World Policy Journal* 6 (Fall 1989): 685–729; Carl J. Migdail, "Mexico's Failing Political System," *Journal of Inter-American Studies and World Affairs* 29 (Fall 1987): 107–23. For more balanced assessments, see John J Bailey and Leopoldo Gómez, "The PRI and Liberalization in Mexico," *Journal of International Affairs,* 43 (Winter 1990): 291–312; Riordan Roett, ed., *The Politics of Economic Liberalization in Mexico* (Boulder, Colo.: Lynne Rienner, forthcoming 1993), both of which combine Mexican and North American interpretations. Another helpful interpretation is Robert A. Pastor, "Post-Revolutionary Mexico: The Salinas Opening" *Journal of Inter-American Studies and World Affairs* 32 (Fall 1990): 1–22. For a brief historical perspective on political modernization, see my "Political Modernization in Mexico: Through a Looking Glass," in *The Evolution of the Mexican Political System,* ed. Jaime Rodríguez (Wilmington, Del.: Scholarly Resources, forthcoming 1992); and Wayne A. Cornelius, "Political Liberalization in an Authoritarian Regime: Mexico, 1976–1985," in *Mexican Politics in Transition,* ed. Judith Gentleman (Boulder, Colo.: Westview Press, 1987), 15–47. Mexican views can be found in Andrea Sánchez et al., eds., *La Renovación política y el sistema electoral mexicano* (Mexico: Porrúa, 1987); and Soledad Loaeza and Rafael Segovia, eds., *La Vida política mexicana en la crisis* (México: Colegio de México, 1987).

For the implications of recent changes for United States-Mexican relations, as they relate to Mexican political development, see Susan Kaufman Purcell, ed., *Mexico in Transition: Implications for U.S. Policy* (New York: Council on Foreign Relations, 1988); Riordan Roett, ed., *Mexico and the United States: Managing the Relationship* (Boulder, Colo.: Westview Press, 1988); Bilateral Commission on the Future of United States-Mexican Relations, *Dimensions of United States-Mexican Relations,* 5 vols. (La Jolla: UCSD Center for U.S.-Mexican Studies, 1989); and Riordan Roett, ed., *Mexico's External Relations in the 1990s* (Boulder, Colo.: Lynne Rienner, 1991), especially part 4. To understand the economic context, see Sidney Weintraub, *A Marriage of Convenience: Relations between Mexico and the United States* (Oxford: Oxford University Press, 1990). An insightful examination, which gets at many of the political and cultural differences, can be found in Robert A. Pastor and Jorge G. Castañeda, *Limits to Friendship: The United States and Mexico* (New York: Knopf, 1988).

Finally, on the issue of human rights and corruption, some literature has recently become available. Stephen D. Morris, *Corruption and Politics in Contem-*

porary Mexico (Tuscaloosa: University of Alabama Press, 1991), is the first book-length work to wrestle with this topic in Mexican politics. Human rights continue to be neglected from an analytical point of view, but the recent publications of international and national human rights organizations provide some firsthand data on the Mexican situation. Among the most useful are the Americas Watch report *Human Rights in Mexico: A Policy of Impunity* (June 1990) and *Unceasing Abuses, Human Rights in Mexico One Year After the Introduction of Reform* (September, 1991), and National Commission for Human Rights, *Second Report* (December 1990–June 1991) and *Third Report* (June–December 1991) (México: National Commission for Human Rights, 1991).

Index